CLIMATIC DROUGHT IMPACT

Oregon's Saline Lake Ecosystem, Lake Abert

Edited by
Frank P. Conte, Ph.D.
Proceedings of the Educational Symposium
held by the High Lakes Aquatic and Alliance Foundation
at Black Butte Ranch, Sisters, Oregon, on April 24–25, 2015

Climatic Drought Impact
Oregon's Saline Lake Ecosystem, Lake Abert

ISBN 10: 1519796331
ISBN 13: 978-1519796332

To order copies, visit **https://www.createspace.com/5931227**
Or contact:
Frank P. Conte
PO Box 170
Camp Sherman, OR 97730
E-mail: **spfclrecon@centurylink.net**

Cover by DeeAnn Glazier, Glazier Graphics & Illustration, Bend, Oregon

Contents

Preface 7

Acknowledgments 9

Part I: Prehistoric Formation of Saline Lake Abert 11

Chapter 1: Prehistoric Evidence in the Formation of Lake Abert
and Its Utilization by Biotic Inhabitants 13
  Prehistoric Evidence in the Formation of Lake Abert 13
  Geography and Geology of the Lower Chewaucan Basin.................... 15

Chapter 2: Human Utilization of Lake Abert's Biological
Inhabitants 17
  Geological Aspects of Lake Abert Shores Revealing
  Archaeological Sites.. 17
  Climatic and Hydrologic Factors Causing Drying of Saline
  Lake Abert 21
  Evidence of Human Utilization of Lake Abert's Biological
  Inhabitants 24
  Indian Burial Grounds Revealed at Lake Abert Shoreline 31

**Part II: Unique Anatomical Structures that Sustain Avian
Migration at Lake Abert 35**

Chapter 1: Morphological and Functional Features of Bird
Species Inhabiting Different Saline Water Environments 37
  Birds Found in Aquatic Environments: Common Morphological
  and Functional Features Needed to Survive 37
  Migratory Birds Living in Open Ocean near Marine Shoreline
  Wetlands 38
  Migratory Birds Living Only in Open Ocean Water 38
  Migratory Birds Living in Hyper-Saline Lake Environments:
  Formation of Toxic Aquatic Systems 42
  North American Salinization of Saline Lakes 48
  Current North American Saline Lakes Having Climatic Impacts
  and Salinization Problems 56
  Physiological Mechanisms Needed by Migratory Birds........................ 56
  Genetic Evolution for Progenitor Stem Cells of Ionocytes 57

Contents *continued*

Chapter 2: Cell Structure of Ionocytes Found in Vertebrate Epithelium 59

The Fish Gill Branchial Epithelial Ionocyte 59

Model of Basal Stem Cell of Ionocyte 60

Differentiation of Progenitor Stem Cells of Fish Ionocytes 61

The Bird Orbital Nasal Gland Epithelial Ionocytes 61

Chapter 3: Migratory Birds and Energy Required for Sustaining Life Cycle 67

Thermal Energy 67

Winter and Summer Grounds at Lake Abert for Migratory Species 67

Power Energy 71

Chemical Energy: Energy in Flight and Energy of Body Functions in Food Sources 75

Part III: Halobionts Inhabiting Salt Lakes: Food Sources of Avian Migratory Birds 79

Chapter 1: Halobionts: Life or Death due to Salinity 81

Halobacteria 81

Halophilic Algae 82

Halophilic Brine Fly (*Ephydra*)—One of the Top Avian Food Sources 83

Halophilic, Branchiopod Crustaceans—Daphnia, Fairy Shrimp, and Brine Shrimp.. 86

Chapter 2: Brine Shrimp Population in Lake Abert: Top Food Source for Migratory Avian Species.. 87

Part IV: Lower Chewaucan River Freshwater Management Issues 91

Chapter 1: Changes in Federal and State Legislation Used by Management Agencies 93

Environmental Changes Requested by Federal Agencies 93

Live Bait Industry—Oregon Desert Brine Shrimp Co......................... 94

Agricultural Lands: Farms and Ranches in the Lower Chewaucan River Basin 101

Health/Medical Industry—Migratory Avian Influenza Virus 115

Water Balance Hydrology of Lower Chewaucan River (Water In = Water Out).. 117

Contents *continued*

**Appendix A: Photographs of Conference Participants
and Audience** 121

**Appendix B: Conference Posters, with Photographs of
Speakers and Abstracts** 127

References 137
    Books.. 137
    Publications and Scientific Journals 138
    Newspaper, Magazine, and Internet Articles........................... 141

Preface

Historical Aspects, Drought Problem

In 2013 the office of the governor of the State of Oregon issued a proclamation that dry conditions, low snowpack, and lack of precipitation have caused natural- and economic-disaster conditions in the southeastern portion of the state. The governor then declared that a state of drought emergency existed for citizens of this southeast region. He then formulated an executive order, No. 14-01, which ordered the Oregon Department of Agriculture, the Department of Water Resources, and the Office of Emergency Management to help mitigate these conditions and provide assistance to the citizens of these affected counties. This executive order urged all citizens of the state to help in these matters.

HLAAF Response to Executive Order No. 14-01

With this background, the scientists and environmentalists who were and are volunteer members of the High Lakes Aquatic Alliance Foundation (HLAAF), and who had worked in the Lower Chewaucan River Basin, instructed the foundation's board members to issue a declaration that HLAAF undertake the goal of preserving Lake Abert's ecosystem.

Lower Chewaucan Basin—A Major Source of Migratory Fowl Populations

The foundation's president, Dr. Frank P. Conte, sent first one letter, and then three subsequent ones, to the Office of the Governor of the State of Oregon, stating the reasons for the critical need to save one of Oregon's most important natural aquatic resources, Lake Abert, from climatic desiccation. We gave the governor our scientific details, which showed him how to avoid the impending desiccation and prevent dissipation of the important migratory birds that use the lake as a sanctuary. These birds require the abundant food resources in this lake for nutrition and energy resources to sustain their continued flight on their migration to Mono Lake, California, and then on to Latin and South America. Unfortunately, the non-profit HLAAF organization received no reply to the four letters as to what action the governor's office had taken or might take in this matter.

Contribution of Lake Abert Avian Species to the Pacific Coast Migratory Bird Populations

As of early 2015, the desiccation of Lake Abert had reached its highest point in 100 years of hydrology. As a result, it has also reached its highest salinity level (>250 mg/l), and all invertebrates in the lake have died, as have all the algae. The lakebed is covered with halobacteria. Most of the migratory bird species have disappeared. To what destinations these birds have gone is open for determination by ornithologists.

At this point, HLAAF decided to make the loss of Oregon's saline lake ecosystem a nationwide story. Its board members suggested that the organization sponsor a workshop on the Lower Chewaucan Basin Aquatic Ecosystem, to take place on Friday, April 24, 2015; and an educational symposium that would focus on Oregon's International Migratory Sanctuary and take place the next day, Saturday, April 25, 2015.

Subsequently it was suggested that HLAAF publish a book expanding on certain topics from the symposium conference, as a means of generating widespread awareness of the saline lake ecosystem issues. The first section of the book would focus on the geographic location of Lake Abert and the geology, hydrology, and paleobiology of the lower Chewaucan River Basin. The second section would cover archaeological evidence of human utilization of the Lake Abert ecosystem. The third section would discuss the anatomical, physiological, and biochemical functions of avian salt glands that permit their inhabiting of saline lakes. The fourth section would focus on the lake's salinization and its malformation with toxic saline systems in salt lakes throughout the world during climatic droughts. The book also would cite evidence of dried California saline lake beds creating human health problems via respiratory inhalation of atmospheric dust and salt crystals. The fifth section would examine the potential of molecular genetic evolution of avian salt-gland stem cell tissue, with the cDNA becoming pathological during embryonic development and causing death. The sixth section would focus on the avian energy requirements for sustaining individual life cycles and maintaining stable body temperatures, in the selection of habitat sites for food sources that provide metabolic energy and growth and development of flight muscles for long-distance flight patterns using stored metabolic energy. The seventh section would be devoted to the description of various halobionic species that inhabit saline lakes, such as halobacteria, halophilic algae, and halophilic invertebrate animals. The eighth, ninth, and tenth sections would cover the freshwater resources needed in management problems that are dependent upon the issuance of new legal permits, or maintaining old permits, for the removal of freshwater from the Lower Chewaucan River Basin for agricultural, forestry, or bait fishery needs. The permit issuance system can sustain or destroy the Lake Abert saline ecosystem.

Therefore, the purpose of this book is to inform Oregon's citizens as to the future of the Lake Abert ecosystem, and also inform citizens of other states as to the ecosystems of their saline lakes. Such lakes are among the many salt lakes throughout the world that provide a most unique habitat for wildlife and natural resources devoted to sustaining the lives of migratory birds.

Acknowledgments

The editor wishes to thank the High Lakes Aquatic Alliance Foundation (HLAAF) Board and members for their support in saving the saline eco-system of Lake Abert. He thanks Harry Wagner for his contributions in support of making the public aware of the Aral Sea drought catastrophe; to Tim Bradley and David Herbst for their investigations of brine fly biology; to Joe Jehl, Jr., for his contributions on the biology of grebes; and to George Keister for his splendid 1992 Oregon Department of Fish and Wildlife (ODFW) report on the saline Lake Abert ecosystem. Also, the editor and his colleagues in HLAAF would be remiss if they did not acknowledge Dr. Ron Larson for his scientific investigations of Lake Abert and his writing and public speaking efforts to save the lake. His referral to the scientific papers published on the pioneering work of hydrology and geochemistry studies by U.S. Geological Survey (USGS) investigators Kenneth Phillips and Steven VanDenburgh, who published an important paper on Lake Abert in 1971, is outstanding. Even today, nearly a half-century later, Dr. Ron Larson considers Phillips' and VanDenburgh's paper one of the major papers on the hydrology and geochemistry of saline lakes. The editor and HLAAF are also grateful for Dr. Rick Pettigrew's writings and lectures on his explorations of early human archaeology. Ms. Helen Schmidling (HLAAF staff writer) wrote a wonderful article, cited in this book, describing Dr. Pettigrew's works on "Prehistoric Life at Lake Abert."

Finally, the editor has had to allow restrictive exemptions for some speakers who could not write a chapter on their topic due to employers' rights or agency obligations. HLAAF granted them their wishes. A few speakers submitted draft copies that underwent review but ultimately did not meet with their complete approval or that of HLAAF. The editor granted these authors the right to revoke their authorship. However, portions of some drafts were approved, and where appropriate in certain topical areas these portions have been cited in the text as Personal Communications (e.g., Author Initials, Pers. Comm. 2015).

PART I
Prehistoric Formation
of Saline Lake Abert

CHAPTER 1

Prehistoric Evidence in the Formation of Lake Abert and Its Utilization by Biotic Inhabitants

by Frank P. Conte

Prehistoric Evidence in the Formation of Lake Abert
The major predictions of when and how the prehistoric Lake Chewaucan (Figure 1) which is located near and based upon shoreline features in Fort Rock region is shown in a drawing from a report by Grayson (1993). A special thanks to Dr. William S. Bowen for the aerial image of Lake Abert (Figure 2). The land that surrounded the ancient Lake Abert was thought to be an arid piece of soil and that during the Pleistocene period of time, large areas that lay south of the lake. However, climatic changes that occurred all over the earth's surface began to modify these arid lands. This was due to the dwindling ice age period, in which rains began to occur and melting snow that had filled the Great Basin began to fill lowlands. One of the lowlands filled was the large region in south central Oregon, along with other Northwestern lands that had an increasing number of new wetlands. It was later shown by early geological investigations that the primitive Lake Chewaucan (Figure 1) was one of these lakes. It was assumed to be quite large: an estimated four hundred-plus miles of lake surface and a depth of more than three hundred feet. This great lake existed some ten thousand years ago. From that period of time until the present, Lake Chewaucan

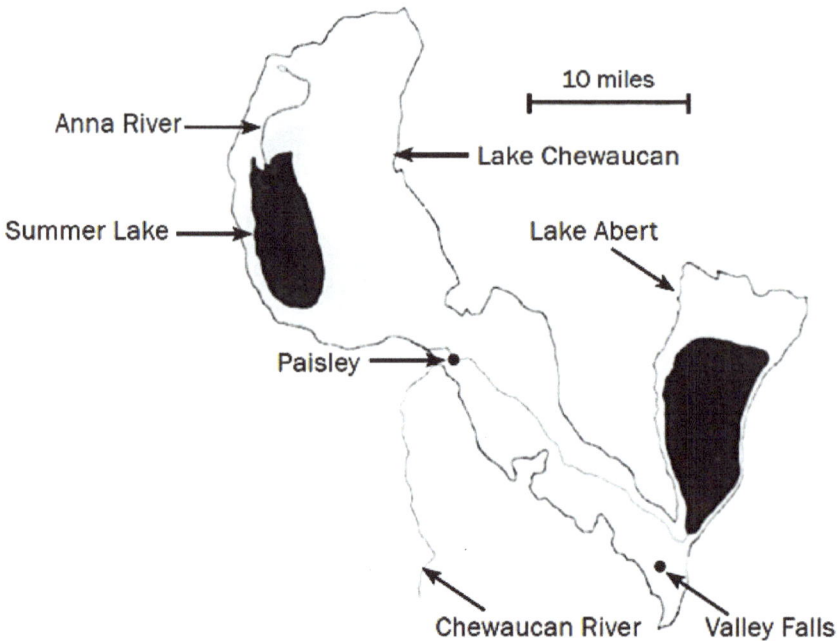

Figure 1. Primitive lakes (grey-shaded areas) giving rise to modern-era lakes (black-shaded areas). (Grayson, 1993)

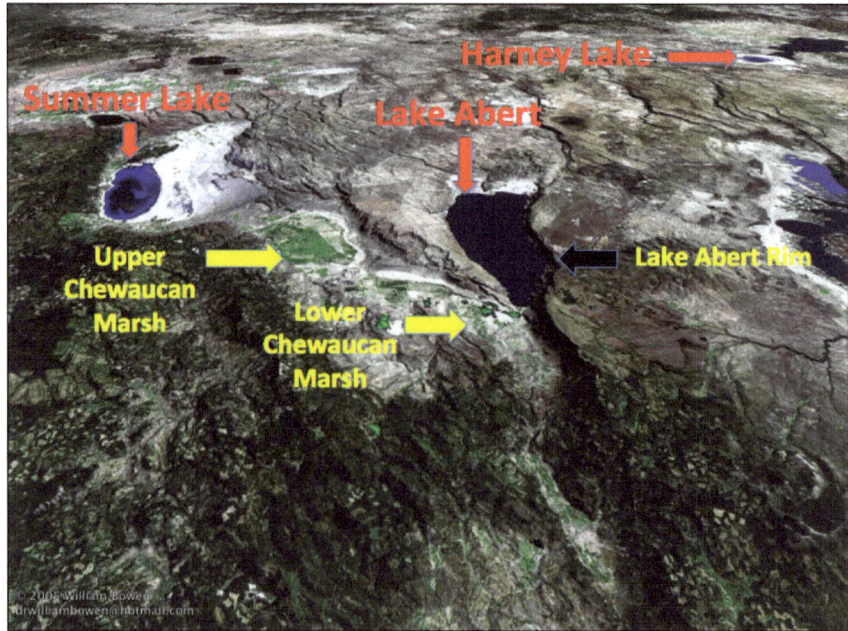

Figure 2. Image of Lake Abert, provided as a courtesy of Dr. William S. Bowen, professor emeritus of geology at the UC-Berkeley, for the *Oregon Atlas of Panoramic Aerial Images* (geodata.csun.edu/ or-panorama.atlas/index.html).

dwindled into many smaller lakes. Today, two fairly large lakes remain: Summer Lake and Abert Lake (Figure 2).

Geography and Geology of the Lower Chewaucan Basin

The Geographic Location of Lake Abert

Lake Abert—or Abert Lake, as it is sometimes called—is located in Lake County in south-central Oregon, not far from the northwest corner of the Great Basin. It has also been included as part of the Great Basin within the physiographic range, which extends from Mexico to Oregon and east to Utah. The lake is eighteen miles east of Paisley, Oregon, and is approximately twenty-five to thirty miles north of Lakeview, Oregon, on U.S. Route 395. This highway is routed along the east shore of the lake for approximately eighteen miles. The highway has several lookout sites that offer views of important aspects of the current biotic inhabitants of the lake, which are described in interpretive signs in these specific regions.

Geological/Hydrological Aspects of Lake Abert

This saline lake is the largest salt lake in the Pacific Northwest and is Oregon's sixth-largest lake, at sixteen miles in length and six miles in width at its maximum recent level. The east side of the lake is bounded by Abert Rim, a steep escarpment that rises over 2,500 feet above the lake's surface. The lake is bordered on the west by a long ridge called Coglan Buttes. However, on the north side of these buttes, the land is referred to as the Coleman Hills. The lake's only year-round source of freshwater, which inflows into the lake's entrance on its southernmost tip, comes from the Chewaucan River. The Chewaucan River has its origin in the territories that form the Lower Chewaucan Basin. The drainages that form this basin area total some eight hundred square miles and are collectively termed "semi-arid."

Much of the precipitation in this area occurs as snowfall during the winter months, and the Chewaucan River system is fed by seasonal snowmelt. The lake's only other source of freshwater is summer thunderstorms, which produce a small amount of runoff from Abert Rim. However, the lake has no outlet for its trapped water. This water, in time, develops a saline condition due to the soils that lie at the lake's bottom (mainly high concentrations of sodium carbonates, alkaline earths, and other mineral rocks). Therefore, these crystalline mineral sources, together with the climatic high summer temperatures that create high evaporation rates of the lake's surface waters, will cause the lake's water to become more saline and alkaline with time. During its 10,000-year existence during the Pleistocene period, the solitary Chewaucan Lake underwent hydrological changes into two types of lakes. It went from a single lake into three smaller lakes: Silver Lake, as a freshwater ecosystem; Summer Lake, as a freshwater/slightly-saline ecosystem;and the terminal lake (Lake Abert), which has become a highly alkaline/saline ecosystem. Lake Abert contains lake water that can be two or three times more saline than the seawater found in the Pacific Ocean off the Oregon Coast (32.5 gm/L).

Watershed of the Two Chewaucan Rivers
The Chewaucan River watershed drains some 645 to 655 square miles containing forests, marsh wetlands, and agricultural ranchlands. It has two main sections of river drainage: (1) Upper Chewaucan River drainage and (2) Lower Chewaucan River drainage. The upper drainage flows easterly from its headwaters throughout the mountainous region until it reaches the town of Paisley, Oregon. Its mountainous regions contain forests of pine, both ponderosa and lodge pole, and groves of aspen and white fir, which are punctuated with tributary springs and creeks lined with many different types of plant life. Below Paisley, the Lower Chewaucan flows through barren valleys and rock terrain covered with marshes of tuffed hairgrasses, redtop sedges, and wild rushes. However, the lower river's flow of freshwater is under state control that allows for the diversion of freshwater as irrigation for pasturelands and marshes, which are surrounded by arid desert soils, before reaching its termination point, the dead-end Lake Abert. This lake is no longer a freshwater lake. Its water has become both highly saline and very alkaline in chemical composition.

CHAPTER 2

Human Utilization of Lake Abert's Biological Inhabitants

by Frank P. Conte

Geological Aspects of Lake Abert Shores Revealing Archaeological Sites

There is a series of sites lying along the biogeographic-ethnic divide close to the town of Paisley, Oregon, that is a rich biotic setting of native aquatic inhabitants. These aquatic organisms would certainly be of interest to people of that very early date. West and south of Lake Abert, extensive bulrush-cattail marshes were supported by the Chewaucan River that continued its flow into the closed basin of Lake Abert. We know from modern times that human residents are fascinated with the prospects of harvesting these productive regions, so the interest by early archaeologists in these Lake Abert locales was to investigate geological sites for cultural features like fire-earth pits, house pits, earth ovens, fish hooks, and projectile points. In addition, cultural habitation residues, such as charred bones, shells, seed, or roots that might be sieved from a fire-earth or trash pit, would give evidence of a people's diet and insight into their hunting and gathering practices. (See H. Schmidling, 2015.)

These tools for establishing the absence or presence of prehistoric inhabitants were used by Dr. Richard Pettigrew and his associates in conducting archaeology research at Lake Abert during 1982–84, under the support of the Department of Transportation. This data was given to HLAAF reporter Helen Schmidling in a discussion with Dr. Pettigrew on his work dealing with "Prehistoric Life at Lake Abert." That discussion was published in the *Nugget Newspaper* on August 26, 2015, and is included here.

> Lake Abert, our state's only saltwater lake, is hardly capable of supporting a human community these days. But Dr. Richard M. Pettigrew of Eugene claims archaeological evidence that human occupation of the lake's margins was once extensive.
>
> Pettigrew says, "Concentrations of prehistoric habitation sites around the lake, on terraces at different elevations, suggest that ancestral Native American hunter-gatherers were able to support surprisingly large populations from the immediately available biotic resources."
>
> This evidence implies that the ecosystem of the lake and the area around it were significantly more productive at a time pre-dating the 19th century. A reasonable hypothesis is that during the Holocene, the lake was relatively fresh and deep, compared to its current condition, and supported a diverse and sizeable biotic population. In other words, enough fresh water, plants, and fish for humans to occupy the region.
>
> Dr. Pettigrew is president and executive director of the Archaeological Legacy Institute. He will speak on the evidence for prehistoric human use of Lake Abert, and its ecosystem implications, at a symposium about Lake Abert on April 25 at the Black Butte Ranch conference rooms.
>
> Like others who've researched Lake Abert, Pettigrew agrees that the high-desert lake seems relatively lifeless, outside of the swarms of brine flies that seasonally cover its alkaline shores. "The rocky landscape, while starkly beautiful, appears inhospitable to human habitation. Historical records mention a Native American presence, but suggest fairly light use.
>
> Pettigrew said, "Imagine the great surprise of archaeologists in the 1970s, while examining a proposed highway project along the lake's margins, when the discovered clear traces of many prehistoric village sites! How could this be? We can't even drink the water!
>
> "We archaeologists never would have been inspired to examine such a place in detail, but the National Historic Preservation Act (NHPA) required serious research during the highway construction planning phase. Whatever our preconceptions, the law demanded that we investigate, so we did.
>
> "After the first inspections," he said, "multiple research phases demonstrated that Lake Abert and its surrounding basin were once home to many people, perhaps thousands of them, who lived in hundreds of

clearly defined houses, many with stone architectural features. People moved their houses up and down the slopes as the lake—which was once much fresher—rose and fell with the changing climate. They lived there until fairly recently. So looks can be deceiving!"

Pettigrew said the first surveys took place in the mid-1970s along Highway 395.

"When I walked the highway route in 1976, I was stunned to find clear traces of a substantial prehistoric society," he said. "The evidence was obvious, and included a staggering array of circular depressions, art motifs, concentrations of all kinds of stone tools, and even the stone-walled ruins of apparent houses."

Subsequent surveys and excavations took place into the 1980s, both along the lakeshore and upstream in the lowlands of the Chewaucan River Basin. The total number of prehistoric sites recorded along the lakeshore and in the basin was 326, including at least 76 village sites.

"Altogether, more than 580 circular features, most of them probably house structures, and 73 rock rings, are recorded," Pettigrew said.

Pettigrew and others have concluded that people have been living in the area for more than 10,000 years, with most of the evidence dating to after 4,000 years ago, and before historic contact. Patterns of site and artifact elevation and relationships with lake terraces indicate fluctuations of lake levels during that period of early human habitation.

There is no currently active program for site protection, beyond the regular Bureau of Land Management (BLM) monitoring of their lands. Tribes were not actively involved in excavations and surveys in the 1970s and '80s, Pettigrew said, but they have become very involved with issues surrounding the River End Ranch and the dam that was built there in 1991, which they claim disturbed prehistoric sites and revealed human bones. "No archaeological excavation took place in relation to dam construction," said Pettigrew, "and this outraged both the archaeologists and the tribes."

The primary source of freshwater for the lake is the Chewaucan River, which was well-known to early fly fisherman and Indians for its adult fish inhabitants, known as "red-band trout," which evolved from ancestral rainbow trout that occurred in the Lahountan Lake region of Nevada. The derivation of *Che-wau-can* is thought to be the Klamath Indian word for *arrowhead* or *duck potato* ("tchua") and a suffix denoting locality ("keni"). The arrowhead is an aquatic plant called *Sagittaria cuneata*, which has an edible starchy rhizome. The lower Chewaucan River is about seventy-five miles long, including two tributaries, Dairy

Lake Abert is shallow, averaging only about five feet deep. Highway 395 runs along the eastern shore of the lake. Like a stretched-out and twisted piece of pie, the lake lies between the brown hills of Coglan Buttes and the precipitous, lichen-covered, basalt cliffs of Abert Rim, which rises several thousand feet above the lake, as shown in Figures 1 and 2 (R. L., Pers. Comm. 2015).

Creek and Elder Creek. Dairy Creek has its origin in a glacial cirque sitting high on the east flank of Gearhart Mountain, which is thirty miles to the west; Elder Creek is in the adjacent mountains near the town of Paisley, Oregon.

The Chewaucan River snakes through lush, green, higher-elevation forests of white fir that give way to drier forests of ponderosa pine and western juniper at lower elevations. There are scattered aspen groves in these forest lands that turn golden in autumn and lie near shorelines of the river as it pursues its final course toward Paisley, where it leaves its canyon and rushes out into the flat basin of Lake Chewaucan, a paleolake that existed about 10,000 years ago (W. R. Tinniswood, 2007).

Climatic and Hydrologic Factors Causing Drying of Saline Lake Abert

All ephemeral salt lakes have no outlet and vary in saltiness at different times of the year. Geologists of yesteryear wanted to have an explanation for this phenomenon. The early geologist Hutchison (1937) began to embrace the idea that "extreme climatic fluctuations could account for the loss of salt from closed basins as is shown by many dry lakes" (W. B. Langbein, Ecological Survey Prof. Paper 412, 1961). Thus was born the concept of climatological evaporation of water vapor as one of the hydrological parameters in water loss or input into closed lakes. In the following years, the evaporation from lake surfaces has been studied extensively and intensively. In Keister's 1992 report, the calculated inflows to Lake Abert were found to be significantly correlated to annual precipitation and flows measured at Paisley (Harding, 1935; Harbeck, et al., 1958; G. Keister, 1992, 2014). This phenomenon of evaporation from lake surfaces in correlation with precipitation probably occurred at Lake Abert from the late 1990s through 2015, with climatic conditions that caused periodic drying up of the lake in 1927, 1930, 1931, 1932, and into the current year, starting with 2014 and 2015.

The details of the impacts on the lake resulting from climatic drought also caught the attention of a staff member of the Oregon Public Broadcasting TV series *Oregon Field Guide*. They produced a segment for the program called "Lake Abert Dries Up." This segment, hosted by Steve Amen, appeared in a episode 2910 of *Oregon Field Guide*, which aired on March 19, 2015. The interviewees featured in the episode—Ron Larson, Joe Eilers, and Keith Kreuz—presented information on the impacts of hydrological changes on the lake's inhabitants. These three interviewees also happen to be members of HLAAF, and they support its goals of saving the lake and its inhabitants. Readers of this book may wish to view this episode of *Oregon Field Guide* since it depicts how the lake shore is increasing in salt deposits, which will eventually enter the atmosphere during winter storms.

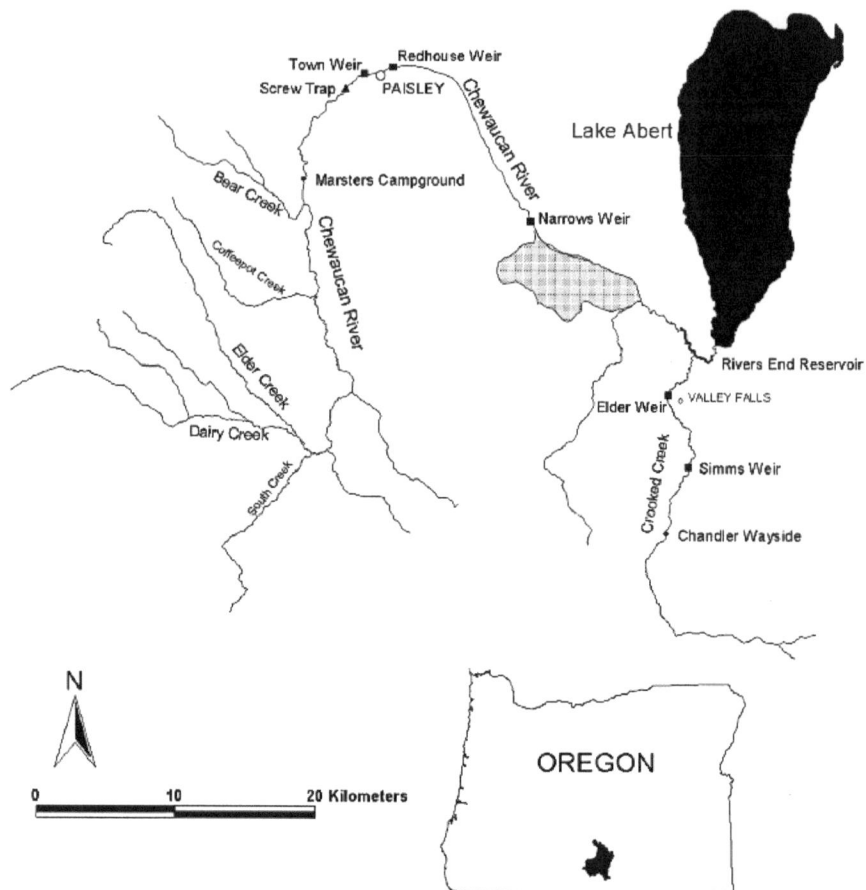

Figure 3. Map of Lower Chewaucan River tributaries and their freshwater flow toward Lake Abert. (Courtesy of W. Tinniswood)

Table 1. Lake Abert Morphometry for the Highest- and Lowest-recorded Lake Surface Elevations, 1916-1965.

	Highest	Lowest
Surface elevation, feet above msl	4,260.5[1]	4,246.6[2]
Surface area, square miles	64	22
Volume, acre-feet	520,000	15,000
Depth, maximum, feet[3]	16	2
Length, miles[3]	16	3
Width, miles[3]	6	1.1

[1] Keister (1992) reported that the "maximum level during the past 100 years was about 4,462 feet above sea level during the summer of 1984." At that elevation, the lake would have covered 65 square miles and contained 560,000 acre-feet of water.

[2] Van Denburgh (1975) indicated that the lake's surface elevation actually fell to 4,245 feet (amsl), probably during the drought years. Approaching dryness at that point, the lake would have covered 12 square miles and contained only 4,000 acre-feet of water.

[3] Estimated values.

Source: Phillips and Van Denburgh (1971); Van Denburgh (1975).

Table 1. Lake surface elevations, as recorded in the lake from 1916 to 1965.

Evidence of Human Utilization of Lake Abert's Biological Inhabitants

For readers interested in more information on archeological science, please see the references for Dr. Richard Pettigrew & Assoc. (1976, 1978, 1980, 1985a, and 1985b).

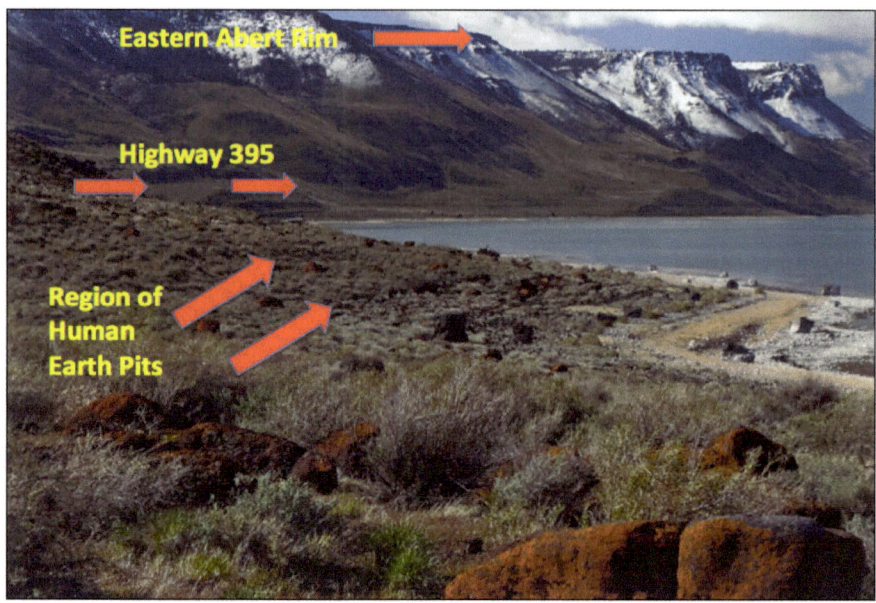

Figure 4. Cultural resources overview of Lake Abert. (See references R. P., 1980; R. P. et al., 1985a; R. P. et al., 1985b)

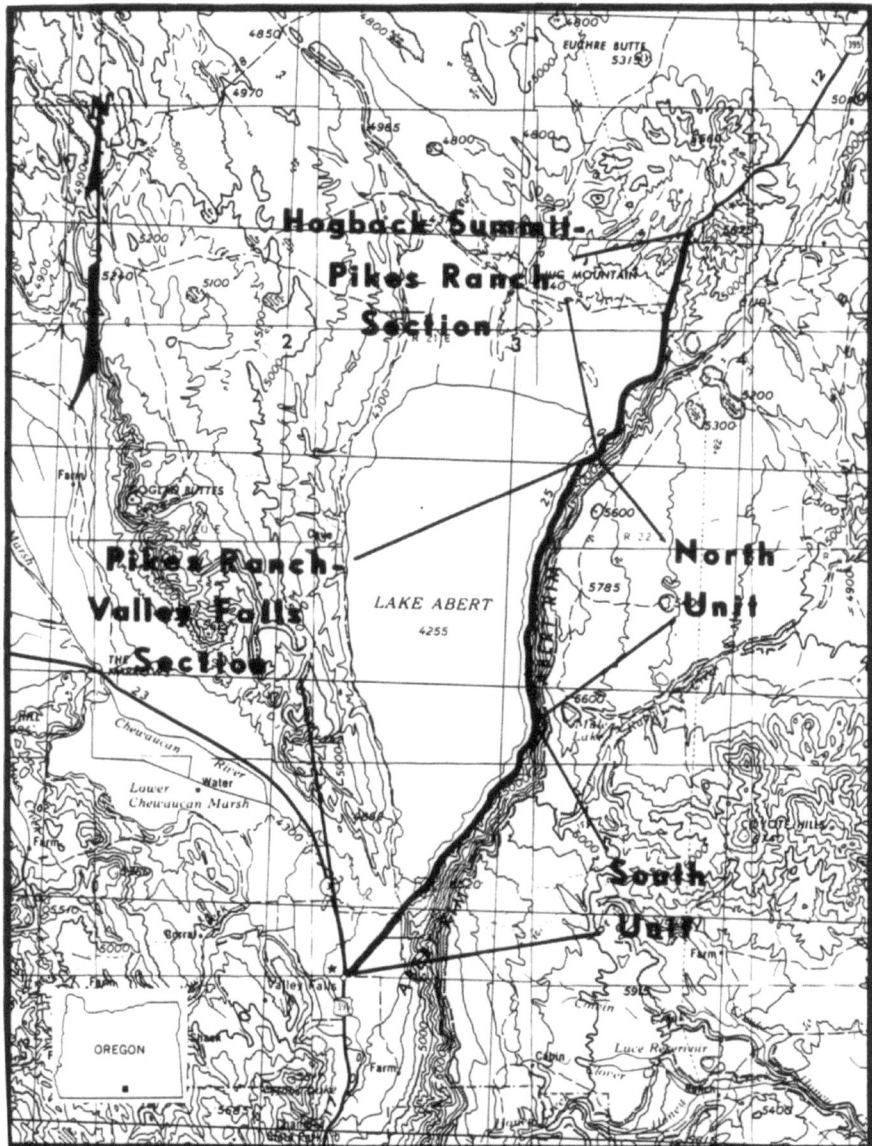

Figure 5. Map of highway sections yielding human activities near Interstate 395 (R. P., Pers. Comm.)

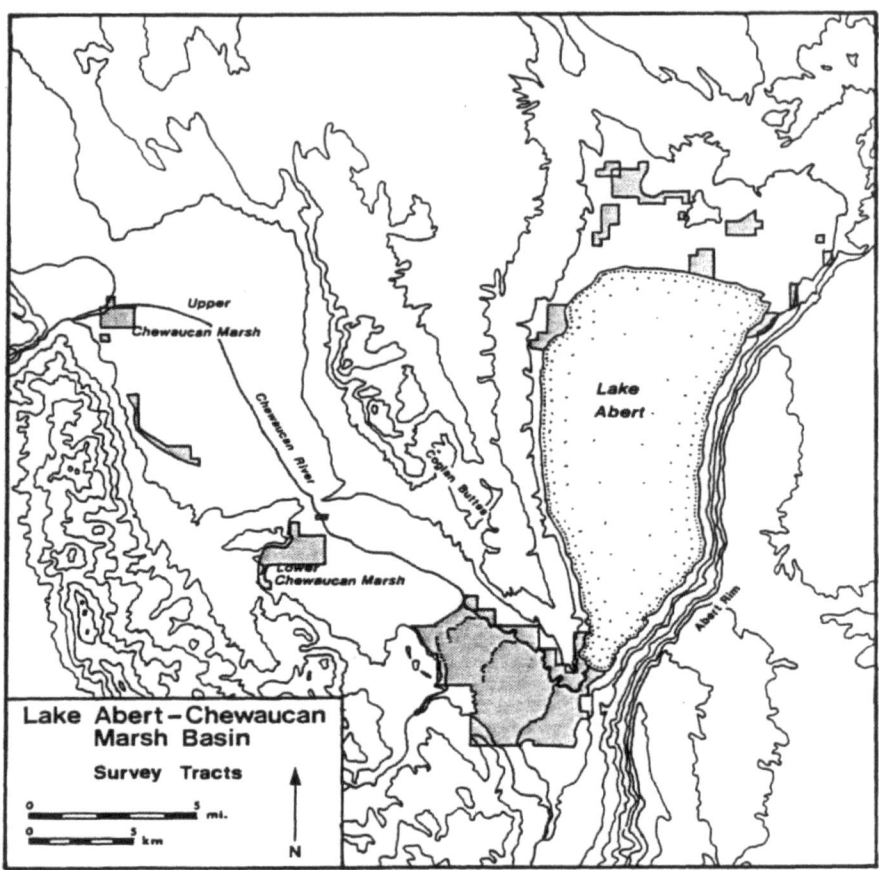

Figure 6. Lake Abert cultural marsh sites. (R. P., Pers. Comm.)

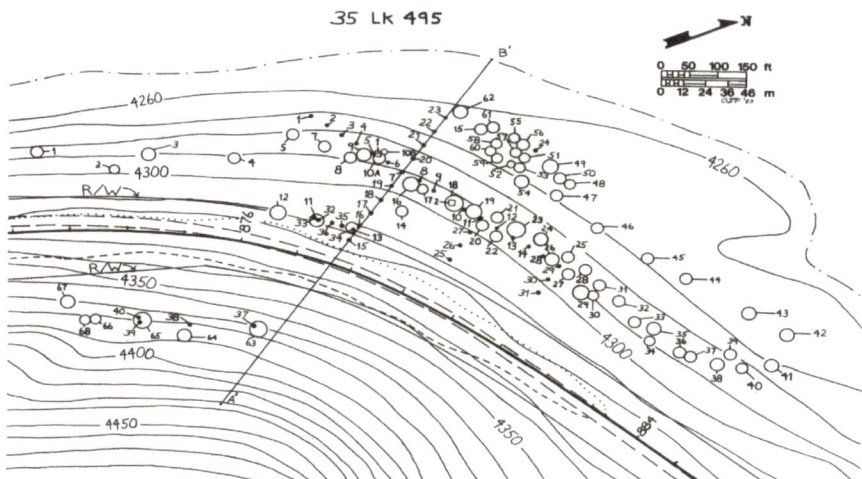

Figure 7. Human earth sites locations (R. P., Pers. Comm.)

Figure 8. Photo of dig site (R. P., Pers. Comm.)

Figure 9. Surface earth site (R. P., Pers. Comm.)

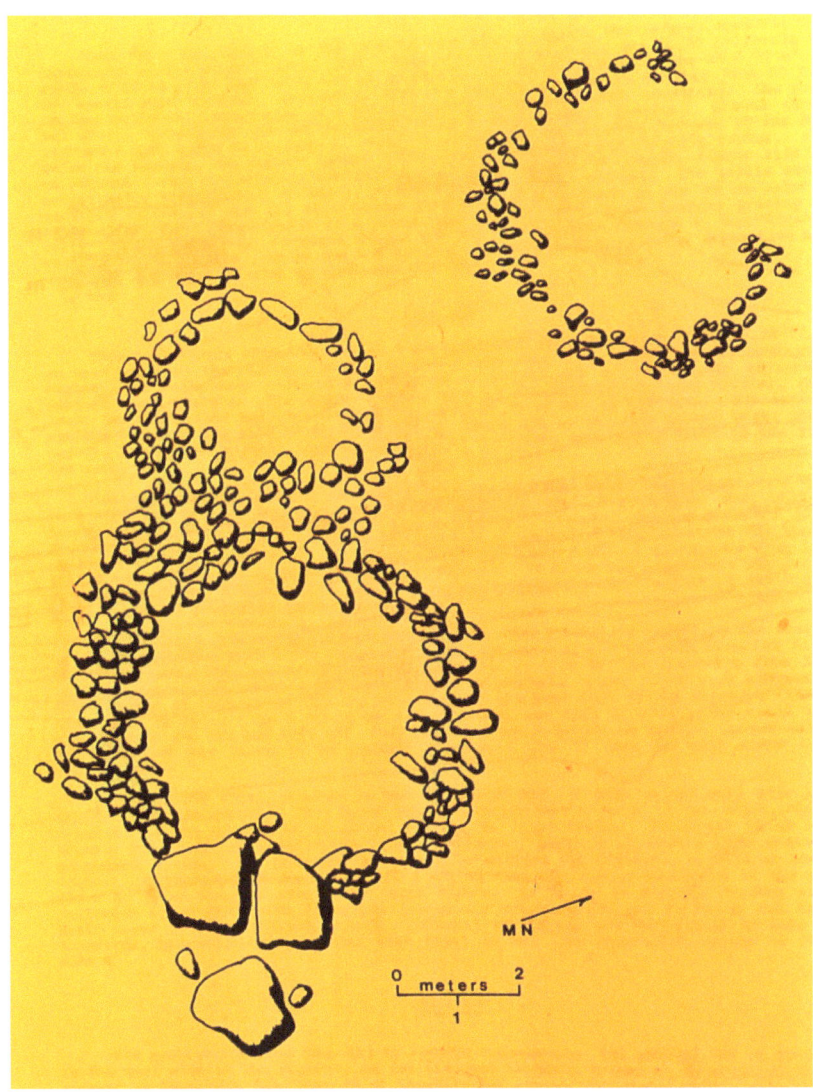

Figure 10. Drawings by scientists of human earth site (R. P., Pers. Comm.)

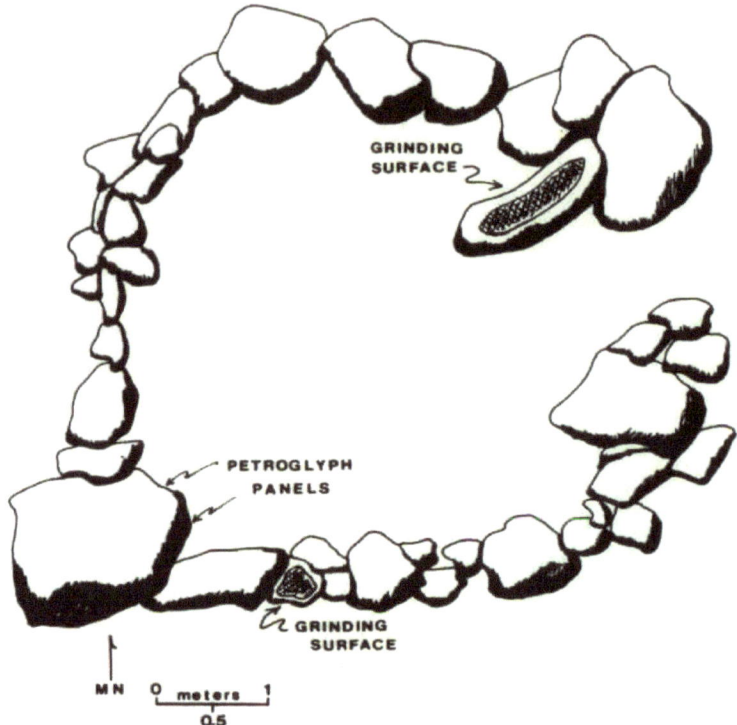

Figure 11. Human earth site drawing by scientists. (R. P., Pers. Comm.)

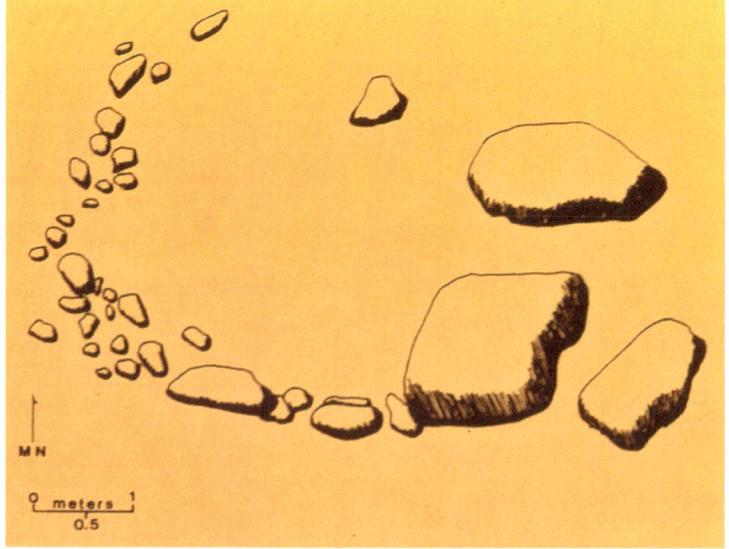

Figure 12. Human earth site drawing by scientists. (R. P., Pers. Comm.)

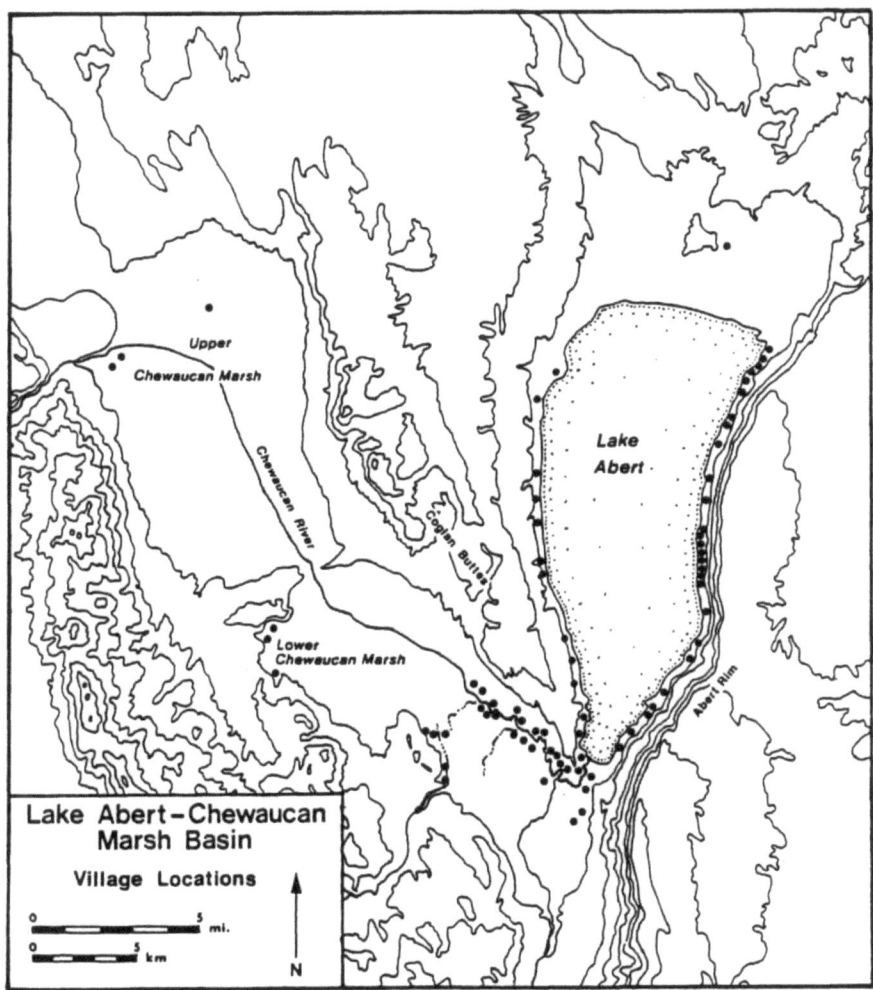

Figure 13. Village locations (R. P., Pers. Comm.)

Indian Burial Grounds Revealed at Lake Abert Shoreline

Evidence was uncovered recently during monitoring of construction on a 1927-era dike on the River's End Ranch property. It was revealed during a search mounted by a member of the Modoc Tribe that the contractor had used burial-ground dirt to fashion a three-foot elevation in the dike. Use of this soil, which contained human skeletons and other artifacts, constituted a violation of the Corps of Engineers compliance rules as well as both Oregon and Lake County laws.

Klamath Tribe Near Remedies Over Disturbed Ancestral Remains
by Cate Montana, Indian Country Tribal Network.com, 2000

The Klamath Tribe is nearing the end of a long legal battle over the disruption of ancestral remains unearthed during a wetland expansion project on the Chewaucan River in 1993.

An agreement with Oliver Spires, owner of River's End Ranch, a private duck hunting club, was reached in February of 2015. A meeting to complete negotiations for government assistance from the Bureau of Land Management and the U.S. Army Corps of Engineers was scheduled with Bob Dodge, attorney for the U.S. Department of Justice, as well as with other government agency representatives in early May of 2015.

"At this point, we're going to sit down to discuss what avenues we have," says tribal attorney Craig Jacobson, an associate with Hobbs, Straus, Dean & Walker in Portland. "What form the settlement will take is, at this point, really completely beyond my capacity to guess."

The River's End Project was a planned expansion of private wetlands that border BLM land on the Chewaucan River. The project included the construction of a dam and a shallow, 700-acre lake.

A private hunting club, River's End Ranch was purchased by Oliver Spires before the dam construction and fill dirt operations began. Original permit applications to the U.S. Army Corps of Engineers under Section 404 of the Clean Water Act included a report completed in 1986, documenting the presence of archaeological sites at the project. A Division of State Lands Waterway Project Permit Review contained a statement by the Oregon State Historic Preservation Office that the "project area is filled with prehistoric sites." The preservation office apparently indicated the project would be acceptable only if efforts were taken to prevent damage to the sites during construction.

In December of 1992, the corps issued Spires a permit under Section 404, and in 1993 construction of the dam began. Although presumably the corps was aware of the archaeological significance of the site, neither the Klamath Tribe nor any other tribe with potential claims in the area was notified.

"They have records that this is a very high-density archaeological area," said Klamath Chairman Alan Foreman. "Our cultural heritage department should have been notified so that we could have a monitor there, so that if something was discovered you could stabilize the site and figure another way to go around it. Ideally that's what should have happened. Apparently somebody didn't do their homework."

It was not until the project was completed in early 1994 that the tribe was notified by an Oregon Burns Paiute Native, who had been invited onto the property, that human bones littered the entire dam surface.

As the river waters rose behind the dam, more remains from Indian graves along the embankments were exposed by water action. In April 1994, BLM agents confirmed that the dam fill used to construct the dam, taken from nearby BLM property, contained human remains. Post facto, the corps suspended the permit issued to Spires. But the damage had been done.

Talks between the tribe, Spires and the corps broke down within the year, long before any out-of-court settlement could be reached. In 1995, the tribe filed suit in district court against Spires as well as the corps and BLM, invoking the Native American Graves Protection and Repatriation Act, the National Historic Preservation Act and the National Environmental Protection Act to seek injunctive and declaratory relief.

"We were asking to be able to sift the dam and the areas where there are remains," Foreman says, "and repatriate whatever exposed remains there are."

The area [needs] stabilizing, bringing in some clean fill and putting up embankments or break waters where it's eroding. "We're trying to get some in-kind services to have the equipment and manpower to do that. Otherwise everything would have to be done by our people by hand. And that won't cover very much."

But to complicate matters, representatives from three other tribes, the Burns Paiute and the Warm Springs tribes from Oregon and the Fort Bidwell Tribe from California, also approached government agencies seeking repatriation rights and restitution. Some remains were turned over to the Fort Bidwell Tribe as well as the Burns Paiute. In January of 1997, the U.S. Department of Justice offered all four tribes a "Global Settlement" of $383,000. It made no further provision for repatriation or stabilization of the site.

The Klamath Tribes' general council turned the settlement down flat, as did the Burns Paiute. Joe Hobbes, vice-chairman of the Klamath Tribes, said the Fort Bidwell Tribe as well as an "individual representative" from the Warm Springs Tribe, took the settlement. As far as the U.S. government was concerned, the case was closed. The U.S. district court in Portland ordered a dismissal against all federal defendants named in the tribe's suit in July 1997.

That decision astounded the Klamath Nation.

"I'm not saying they (other tribes) don't have interest there," Hobbes says. "Some of those bones are Paiute bones. The Paiutes were wanderers. They never stayed in one specific area. "But…we gave up 22 million acres of land in order to get a 2.2-million-acre reservation, and that River's End Ranch is right inside of our ceded land boundaries. We're the only ones who have proven direct linear descendancy. No other tribe's proven that. That's why we never could figure out why they're (the government) giving to those other people what is ours."

To add insult to injury, the court issued an order awarding attorney fees to the private defendants—$40,000, to be paid by the Klamath Tribes.

A series of appeals by the tribe and proposed settlements followed. In June 1999, Klamath tribal council members, after numerous requests, were given permission by Spires to personally inspect the site for the first time.

Despite the settlement with Warm Springs and Fort Bidwell, no reclamation work had been done on the site. Witnessing the exposed skeletons and human remains scattering the dam and shores was a tremendous shock. Hobbes and Foreman both recall their dismay at discovering skeletal remains and several skulls as they walked along the shoreline.

"There was one complete skeleton and three skulls," says Hobbes. "When I walked up on that dam, I found some pieces of human bone there, little chunks that their machinery had broken up. We called Spires to come see and he said, 'I don't want to hear it.' He wouldn't even go over there."

As disturbing as the discoveries were, the site visit opened doors to continued talks. Since that time, Foreman said, Spires has been willing to cooperate with tribal members. A final settlement signed between the tribes and Spires in February permits the tribes to remove remains from the dam and back-fill with clean dirt. Additional fill will be used to cover exposed remains along the shoreline, and breakwaters will be built. The agreement is renewable after two years, and is everything the tribe hoped for since talks began in 1994. Work will begin in the fall when lower water levels won't impede the process.

In the meantime, the tribe hopes a settlement with the U.S. government will provide the equipment and money that the project will require.

PART II
Unique Anatomical Structures that Sustain Avian Migration at Lake Abert

CHAPTER 1

Morphological and Functional Features of Bird Species Inhabiting Different Saline Water Environments

by Frank P. Conte

Birds Found in Aquatic Environments: Common Morphological and Functional Features Needed to Survive

Why do birds drink water? It has been found that all bird species need water to keep their internal body organs covered with safe, healthy fluid. This internal-body water must be kept at a certain volume and must contain unique concentrations of electrolytes in the vascular blood compartment and the lymphatic compartment. The heart and blood vessels circulate blood fluid that contains all the food and respiratory ingredients needed to sustain the birds' lifestyles. The lymphatic system primarily aids in sustaining the cellular health system by removing toxic agents, such as bacteria and viruses. The renal organ handles toxic metabolites, such as the nitrogen-containing urea or excess electrolytes ($Na+$, $K^{+, Cl-, SO}4$, etc.).

This blood-regulatory process is referred to as osmotic and ionic fluid regulation. In terrestrial birds, the kidney or renal system is the most important organ in maintaining this critical process of sustaining the composition of the circulating blood and lymph fluids. Other organs can play a minor role, such as the digestive tract and lymph nodes (see Conte, F. P., P. A. Conte, and C. P. Conte, 2014, *The New Biological Secrets of Salt: Its Diversity in Organisms and Its Impact on Humans*, on Amazon.com).

Migratory Birds Living in Open Ocean near Marine Shoreline Wetlands

Do all of these bird species utilize the same kind of organs to sustain identical osmotic and ionic fluid balance? Aquatic birds that live in open-ocean seawater but near the seashore have sporadic access to freshwater. These wetlands in the adjacent seashore regions can have a variety of freshwater sources, such as springs, creeks, potholes etc. Therefore, these marine birds have access to freshwater for sustaining their blood composition and volume. These internal fluids do not becomes metabolic or salt loaded, or what is referred to as becoming hypertonic. The renal system with the kidney still remains competent in maintaining the volume and composition of blood fluid.

Migratory Birds Living Only in Open Ocean Water

Should the living habitat become either completely open ocean, with no shoreline within short flying distance, or dried-up due to a climatic drought and transformed into a inland playa containing only a small, isolated salt lake as its remaining aquatic region, can the kidney of the migratory bird sustain the osmotic and ionic regulation of blood volume and the excretory system?

Discovery of a New Salt Regulating System: The Orbital Salt Gland

In 1960, K. Schmidt-Nielsen and coworkers began investigating the marine sea gull, which lives its entire life in the open ocean. How was the bird able to regulate its blood and other internal fluid compartments? Did the kidney do the complete job? They found that the kidney in this case becomes a dysfunctional organ (K. Schmidt-Nielsen et al., 1960); the renal system requires additional help. Since this aquatic region usually is saline, and there is no freshwater resource to reduce this increased salinization due to evaporation of freshwater-vapor from the ocean's seawater, it becomes what is referred to as being hyper-saline. A new organ that can aid the renal kidney system in sustaining osmotic and ionic balance in birds is called an extra-renal organ or salt gland.

Extra-renal organs thus became a new and important regulating structure. These organs, which contain mixed tissues, previously had various names, primarily due to their cell structure, such as chloride cells, mitochondria-rich cells, etc. Ultimately it was proposed that these structures be called salt glands containing ionocytes (Conte, F. P., 1980, in book by Lahloo).

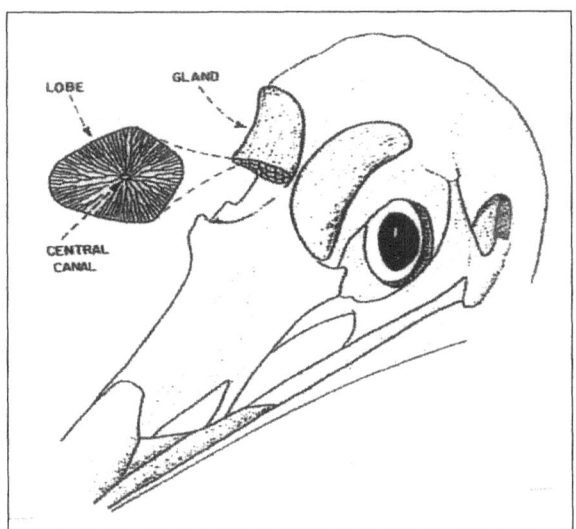

Figure 14. The skull of an adult gull taken from the sea was dissected to find where the orbital salt gland was located. The skull showed two shallow depressions lying above the eye that contained a multi-lobed organ containing epithelial tissue. (Schmidt-Nielsen photos, 1960)

Schmidt-Nielsen and associates (1997) were the first investigators using marine gulls that drink seawater to show that osmotic and ionic water balance could be obtained by way of the orbital nasal gland secreting excess sodium chloride solution being eliminated from ducts emptying into their beaks.

Biochemistry and Molecular Biology of Extra-Renal Organ Now Known as Avian Salt Gland

This extra-renal organ has now become a new—and the most important—ion-regulating structure for saline-adapted aquatic birds. These salt glands contain a mixture of cell types: some cells have the structures for handling the excretion of sodium ions or chloride ions; others make large quantities of the energy-rich compound ATP, are rich in mitochondria, and are termed MR cells, etc. It was finally proposed that all of these cells, which belong to the epithelial cell family, be termed as epithlelial ionocytes. (Conte, 1980). This came about by the biochemical demonstration that several types of tear, nasal, or salivary glands containing these epithelial cells had all types of enzymatic ion pumps. Therefore, they should be termed as being secretory ionocytes. These epithelial cells are responsible for the synthesis of the ion-pump protein structures and required sensory-messages of miRNA coming from cDNA areas, initiating cellular differentiation of adult stem cells containing the genes needed for making ion-pump structure and their location on the cell surfaces (see Chapter 4 for more detail).

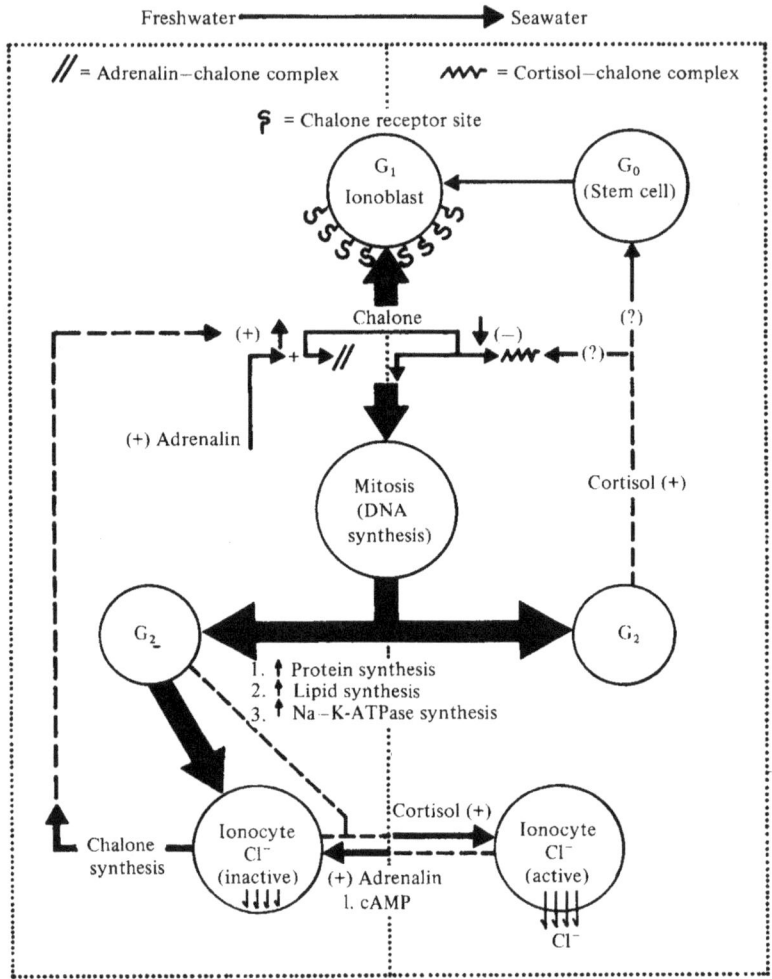

Figure 15. Lahlou's hypothetical schematic of ionocyte formation.

Alternative to Salt Gland: Discovery of Avian Beak Skull Grooves and Tongue Growths as a Water Sequestration System

But not all aquatic birds use salt glands for this purpose, as was shown by Mahoney and Jehl (1987), who investigated the aquatic grebes living in Great Basin salt lakes. They found that grebes were able to "drink" saltwater via the special grooves in the skeletal bone structure of the beak tips. These grooves create adhesive and capillary action with the soft surface-tissue and muscular tissue of the tongue for holding food particles and allow the external saltwater flow to pass as fluid into the pharynx, out of the bill, and

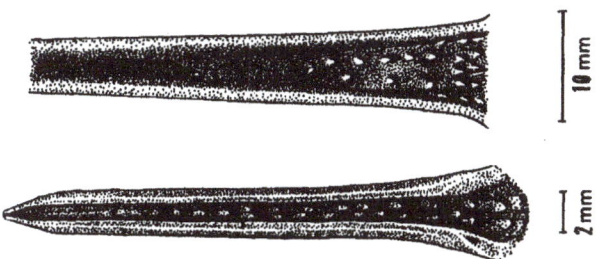

Figure 16. Beak structural grooves in grebes. (Mahoney and Jehl Jr.)

Figure 17. Aquatic grebes at Lake Abert. (Courtesy of A. Henry)

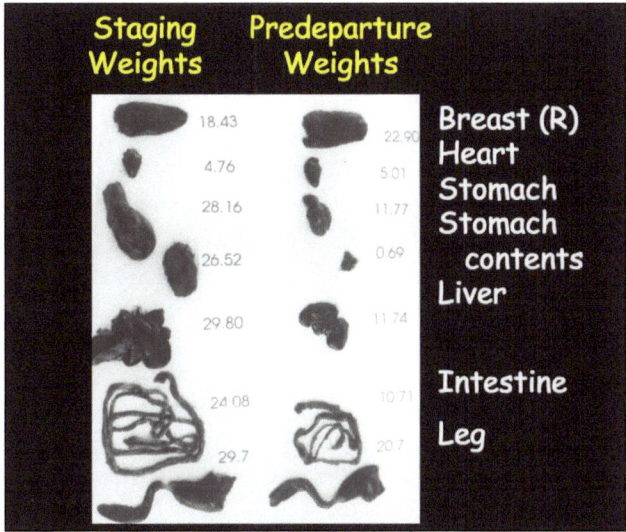

Figure 18. Grebe anatomy dissection. (Courtesy of A. Henry)

back into the environment. Therefore, this salt does not enter the internal body (see Figure 16).

Do other species of birds use this method of salt elimination? Do young chicks, during the development of their salt glands, use the beak system? These questions still need to be investigated to provide a better understanding of how nature has developed this extra-renal system.

Migratory Birds Living in Hyper-Saline Lake Environments: Formation of Toxic Aquatic Systems

Salinization of Wetland Lakes: Normal Geological Processes that Can Cause Toxic Salinity

Climatic drought is caused by changes in the atmospheric conditions that cover various types of global regions. If the atmospheric clouds and wind conditions cause a reduction in the formation of rain or snow particles deposited in regions of mountainous basins and agricultural lands, these regions become dehydrated. They use headwaters of streams that feed freshwater into lakes and other wetlands or other types of soil reservoirs, and the solute concentration in these water bodies may undergo major changes in saline composition. A dramatic example of how such a change could occur was seen in the San Joaquin Valley in California, where the Kesterson National Wildlife Refuge contained a wetlands aquatic refuge with suitable habitat but poor water-solute quality composition. It attracted marine shore birds of the wading type, which used the pond as a food source and breeding ground to substitute for the loss of other aquatic habitats. Since Kesterson Aquatic Refuge was supplied with freshwater from an irrigation water drainage system, it was considered safe from over-salinization of sodium salts coming from the ocean or soils. Unfortunately, the refuge's management was unaware of the contaminants that accompanied the rainwater, supplying the habitat with other ions such as selenium and heavy metals, from the recycling of drain water and soil-leaching waters. The accumulation of these solutes caused massive toxicity to the birds' fertile eggs and resulted in major anatomical malformations, resulting in the death of newborn chicks (D. J. Hoffman, 2002). Since becoming aware of these toxic factors, the Kesterson Refuge has filled the wetland with uncontaminated soil, in order to displace breeding birds to wetlands of better water quality. Therefore, wildlife environmentalists have suggested that there are equally dangerous, and potentially much more widespread, mechanisms of water-quality reduction, via alterations in the salinization of hyper-saline lakes, be they ephemeral or terminal.

July - September, 1989   October 5, 2008

Figures 19 through 22: Aral Sea photos. (Courtesy of H. H. Wagner, ret. ODFW [Pers. Comm.] and Aral Sea Crisis, www.Columbia.Edu)

Salinization in World Agricultural Lands

Salt lakes occur worldwide in dry regions, and it is common to find agricultural farms growing plants using freshwater diversions from rivers inflowing to these lakes on rich soil lying on top of these arid lands. For example, in North Africa, Southern and Western Asia, and Central Australia, you will find numerous plantations using freshwater to grow their crops.

The Aral Sea, and the entire Aral Sea Basin, has achieved worldwide notoriety because it is nearly completely dried up. At one time, it was the fourth-largest lake in the world, fed by two major rivers, the Syrdarya and Amudarya, which formed the natural border between Kazakhstan and Uzbekistan (photos by Michael Shamshidov, OrexCA.comm; Aral Sea Ecological Disaster—Uzbeki, George Kowrounes).

The International Geographical Union singled out the Aral Basin in the early 1990s as one of the earth's most critical environmental zones. In fact,

Figure 20. Boats in Aral Sea crisis.

Figure 21. Aral Sea fishing boat abandoned after grounding on playa.

Figure 22. Aral Sea salt bed, growing food for camels.

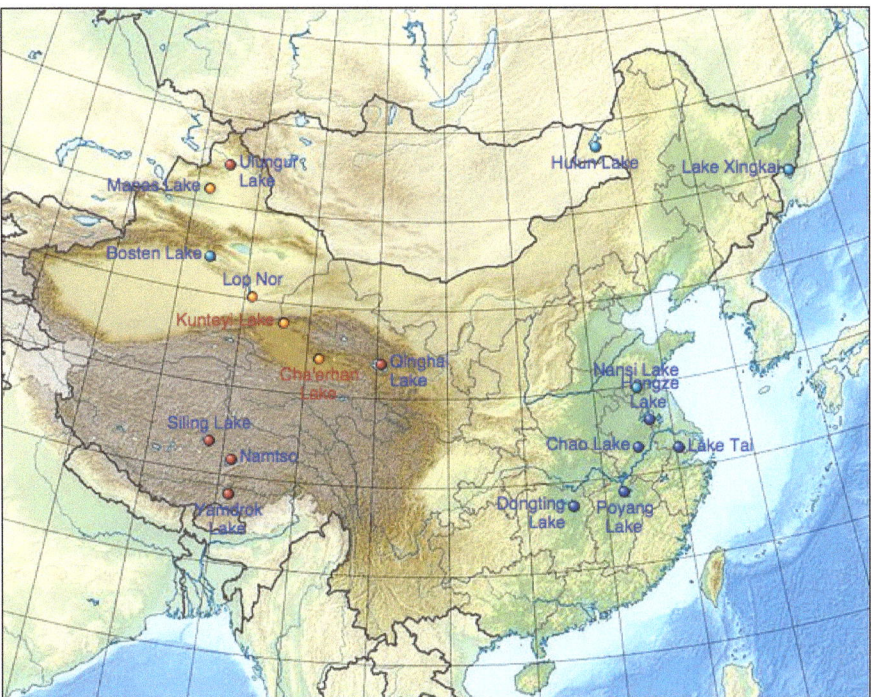

Figure 23. The five largest saltwater and freshwater lakes in China. Red = largest five; orange = dried ones; blue = traditional five largest lakes; cyan = those that are supposed to be in the largest five. (Courtesy of Wikipedia: https://en.wikipedia.org/wiki/List_of_lakes_of_China)

Figure 24. Qinghai Lake, the largest saline lake in China.

Figure 25. Tuz Saline Lake, once the largest salt lake in Europe and the second-largest in Turkey, faces drying up because of local water policies.

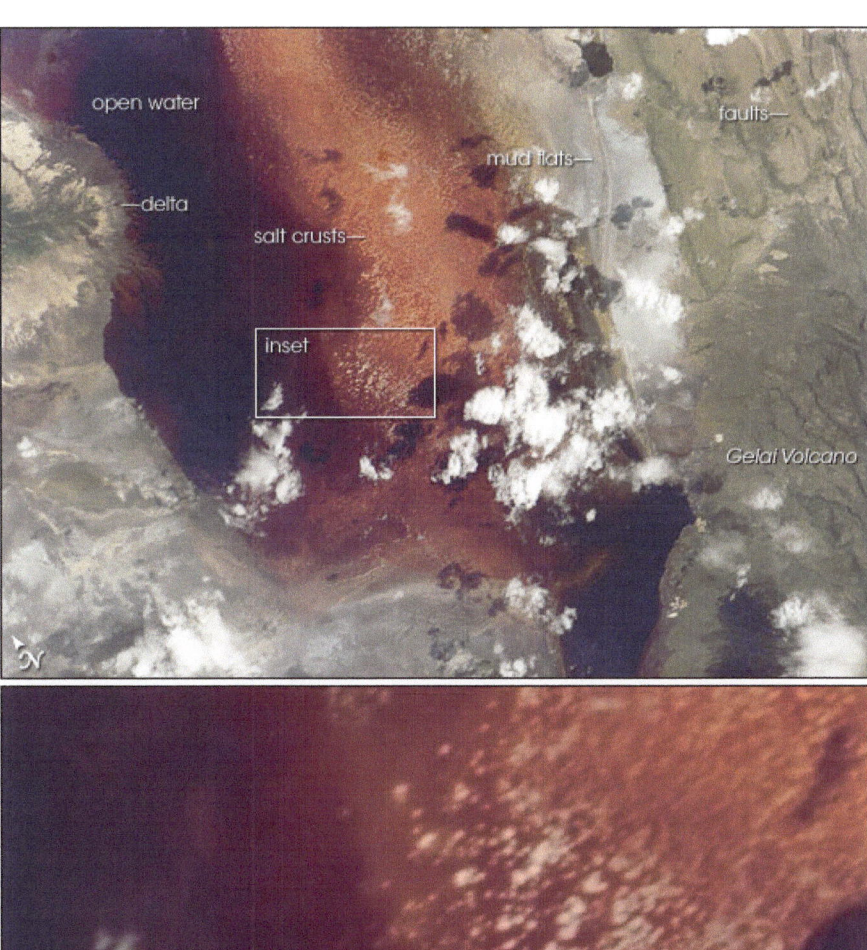

Figure 26. Natron Saline Lake, Tansania. The lake is close to the Kenyan border and is in the Gregory Rift, which is the eastern branch of the East African Rift. The lake falls within the Lake Natron Basin, a Ramsar Site wetland of international significance. The red color of the lake is characteristic of very high evaporation rates. As the water evaporates during the dry season, the salinity levels increase to the point where salt-loving organisms can thrive. Such halophile organisms include cyanobacteria, which is described in later chapter of halobions.

Figure 27. Natron Saline Lake is the only regular breeding area in East Africa for the 2.5 million lesser flamingoes, whose status of "near threatened" results from their dependence on this one location. When salinity increases, so do cyanobacteria, and the lake can also support more nests. These flamingoes, the single largest flock in East Africa, gather along nearby saline lakes to feed on spirulina (a blue-green algae with red pigments). Lake Natron is a safe breeding location because its caustic environment is a barrier against predators trying to reach their nests on seasonally forming evaporite islands. Greater flamingoes also breed on the mud flats.

a scientific group of salt lake experts (including this author, F. P. Conte, 1995) were escorted to the basin under Russian Professor A. Aladin's leadership, opening Russia's doors to the basin in an effort to determine how to manage the two rivers in the interest of saving this important saline lake. At that time, as can be seen in the following photographs, the Aral Sea sustained a natural fishery economy and supported many businesses for the two countries. As can be seen today, this fourth-largest lake is a global environmental and humanitarian tragedy.

North American Salinization of Saline Lakes

In North America, salinity from agricultural rainwater, surface flow, and subterranean flow is a problem in most of the western states' inland waters, to some extent, and in some mid-eastern basins. In the arid American western states, many factors tend to increase the salinity of inland water habitats. Increased urban uses of riverine waters decrease the amount and quality of water available for dead-end lakes. Effects of such decreases include the drying up of some of the dead-end habitats, reducing freshwater availability in the landscape. In addition, the remaining wetland habitats become shallower and more saline, through the evaporative loss of water vapor, making the remaining water saltier.

One of the best examples of this is the water diversion of the eastern Sierra Nevada snowmelt into the Owens Valley Basin.

Desiccated Owens Saline Lake Bed Atmospheric Dust/ Salt Problems

Owens Lake was a terminal lake to Owens River throughout prehistoric times and held water for at least the last half-century. During this period and between drought conditions, it served residents of the area with agricultural diversions. This human water usage enabled the City Los Angeles water department to purchase land along the Owens River. In this manner, the city in 1913 began obtaining water-diversion permits for the Owens River, enabling the city to create and construct a nearby freshwater reservoir for residents. During the last quarter of a century, this diversion resulted in the depletion of all the Owens

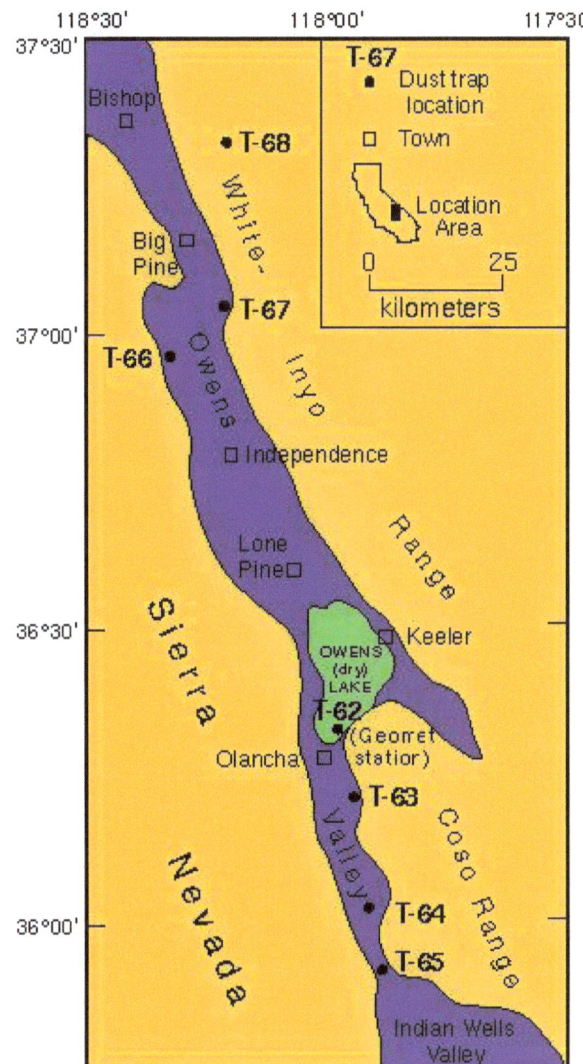

Figure 28. Owens Basin—Dry Owens Lake Salt Bed/Gordo Cerro Salt Mine. (Map from California's Pink Salt Lakes—Wayne's Word)

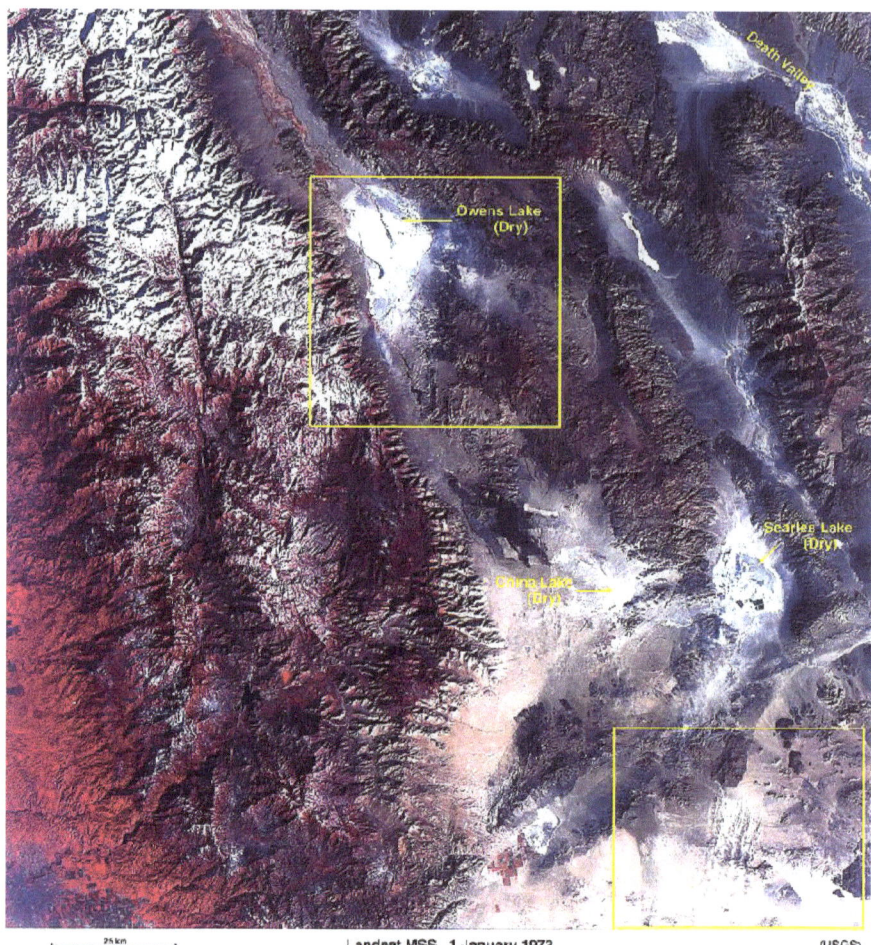

Figure 29. Dust storms in Owens Valley. A Landsat MSS image of dust clouds in the region east of the southern Sierra Nevada. Yellow squares show enlarged areas of Figures 30 and 31. (Images courtesy of P. Chavez, USGS)

River and Owens Lake water, and eventually the lake dried up. Later, this diversion caused a reduction of water inflow into Mono Lake, thereby increasing its salinity. Eventually the lake's salinity tripled, leading to a drastic reduction in its use by thousands of migratory ducks. In addition, the shoreline of Mono Lake became caked with dried sodium-chloride salt crystals. During late summer and between inter–winter months, the wind blew across the salt beds and vaporized the soil into dust storms of salt and soil particles, which filled the lake's surrounding atmosphere with dust plumes that could easily dissipate into other geographic regions (J. S. Reid, et al. 1994). These fine-grained, alkaline dust particles have an estimated

Figure 30. Two dust plumes originating from Owens (dry) Lake. The southern plume is moving south past Olancha; the northern plume is moving southeast over Keeler. (Courtesy of P. Chavez, USGS)

size of less than ten microns (PM10 aerosol dust particle), and the lake produces an estimated seven to eight million metric tons of these particles per year (Gill and Gillette, 1996; M. Reheis, 2015 in press).

Owens Lake Health Hazard Due to Dust Storm Inhalation
Because these dust particles are so small, they can be inhaled deeply into the human respiratory tract and cause serious health problems (Reid, et al. 1994) Therefore, the State of California and the United States Federal Health Service have had to monitor this health hazard. Since 2001, the city of Los Angeles has been required to mitigate the dust storms each year. As a consequence, these toxic storms caused the city of Reno, Nevada, to mitigate their impacts.

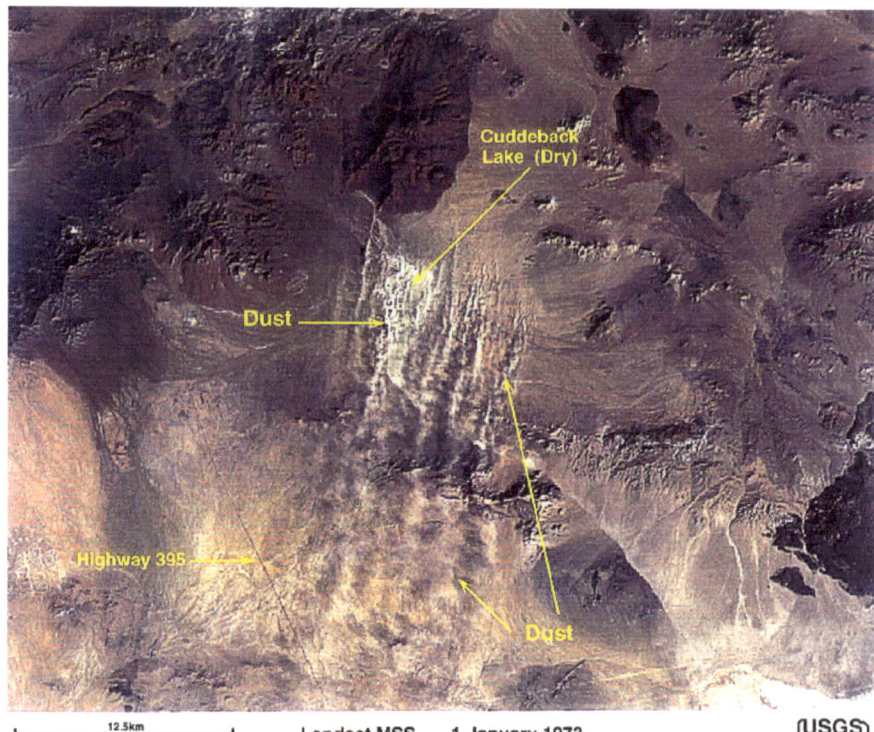

12.5km   Landsat MSS   1 January 1973   (USGS)

Figure 31. Dust plumes streaming south from the area of Cuddeback (dry) Lake. Note that many of the individual plumes are being emitted from alluvium, rather than from the playa (Reheis and Kihl, 1995).

California Pink Salt Lake—Dried Owens Salt Lake (M. C. Reheis [in press])

The Owens Lake Legal Investigation

The legal investigations showed that the salt particles in PM10 initiated the electrical breakdown of electronic instruments used in recreational machines—within gambling pavilions. They lost function in addition to the loss of patrons, due to the potential health hazard. What followed was a legal suit brought by the State of Nevada against the State of California, to solve the problem of water diversion from Owens River by the Los Angeles Water District (LAWD). The solution ultimately came through a Supreme Court ruling that the Public Trust Doctrine must be applied to the management of this freshwater system. The details of the management ruling are found in a book edited by Duncan Patten (1987), titled *The Mono Basin Ecosystem: Effects of Changing Lake Level* (National Academy Press, Washington, D.C.)

Figure 32. Owens Basin. A strange phenomenon caused by red halobacteria. Owens Lake bed along Highway 395 south of Lone Pine, with Sierra Nevada in the distance. (Photo by W. P. Armstrong, 2006)

Figure 33. Solar evaporation ponds of the abandoned chemical plant located at the northwest end of Owens Lake. Note the vivid red of the salt-loving bacteria.

Figure 34. Collecting brine from the chemical plant in Trona, California.

Figure 35. The shovel is stuck into saturated brine in Owens Lake containing red halobacteria and a homogeneous population of the green algae *Dunaliella salina* (Figure 36).

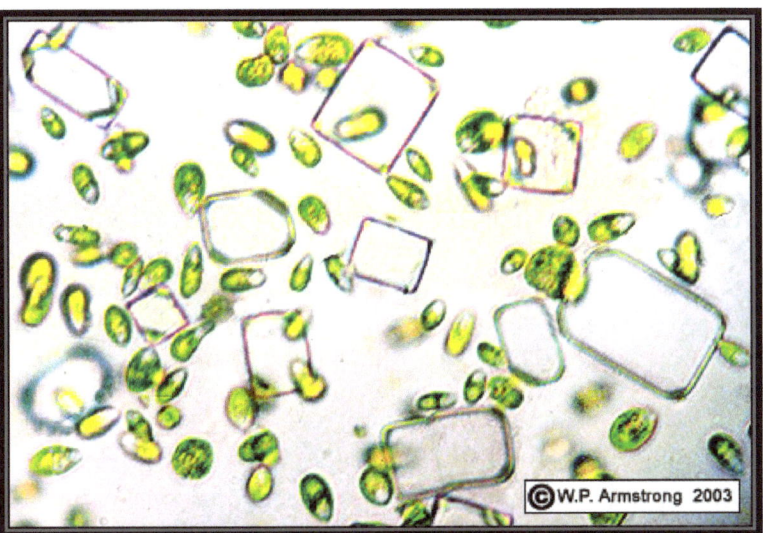

Figure 36. Microscopic view of the green algae *Dunaliella salina*.

Current North American Saline Lakes Having Climatic Impacts and Salinization Problems

At one time, targeting lake salinities that are compatible for maintaining bird populations was thought to be a simple matter. Numerous environmental geologists have charted and measured water qualities of existing U.S. wetland habitats throughout our state lands. They indicate which bird species inhabit this water, along with their numbers. They have correlated bird populations with chemical definitions of salt solutions that do not detail the solutes that comprise this general physical terminology. Rubega and Robinson (1996) have shown that shorebirds that inhabit wetlands are having emerging issues correlated not only with the chemistry of the waters but with other biological problems. Table 2 has a simplified version for the classification of wetlands with respect to salinity adaptations.

Table 2. Classification of types of water (wetlands) with respect to salinity. Modified from Cowardin *et al.* (1979).

Conductivity (S/cm)	Classification
> 60,000	Hypersaline
45,000 - 60,000	Eusaline
30,000 - 45,000	Polysaline
8,000 - 30,000	Mesosaline
800 - 8,000	Oligosaline
< 800	Fresh

Physiological Mechanisms Needed by Migratory Birds

But what is most important (and is lacking in Table 2) are the biological factors that can be deduced from the survival of organisms living in these saline lakes. They are the organisms used by birds as their food sources, needed for the growth and development of young bird chicks and for adult migrations in completion of their reproductive lifecycles. Physiological measurements of their osmotic and ionic concentrations as a correlative index is lacking. It is essential for this information be added to the nomenclature in this simple tabulation. Setting target salinities depends not only on a better biological understanding of how effects accrue in particular species, but also upon policy decisions about acceptable effects for species richness at the population level. The best example of how this system of targeting salinities should work has been developed in England. At Havergate Island, the production of avocet chicks depended upon measuring the freshwater inflows together with monitoring the chicks' physiological responses to salinities. The volumes of freshwater inflows needed in the constructed lagoons to achieve the desired salinities for maximum growth of chicks were both large and costly. Therefore, for arid lands in the western states, this approach is very limited in application, as is known by many managers overseeing terminal lake habitats that rely on basin freshwater river inflow.

At the ecosystem level, other types of biological investigations have shown how salinization alters ecosystems by the destruction and/or alteration of populations of insect species, halophytic plant or algal species, and halo-microbial or viral communities. These species are able to continue living in these waters through modifications in their physiological mechanisms, allowing them to adapt to changes in salinity. All of these biotic species possess osmotic and ionic regulations that make wetland habitats sustainable.

Genetic Evolution for Progenitor Stem Cells of Ionocytes

It is in the same manner that the physiological system of salt glands that evolved from the modification of tear ducts in orbital nasal glands found in certain species of avian birds became extra-renal salt secretory structures. It has taken nature thousand of years to evolve and cope with these changes in internal water composition, as seen in freshwater birds with only the terrestrial type of tetrapod kidney nephron for survival. Thus, during the salinity changes in the prehistoric period where freshwater birds' livelihoods were being altered, the adult bird had to genetically develop a new type of cell to osmotically handle the load of new solutes entering their vascular systems. The existing kidney system could not excrete these solutes. Therefore, the adult bird had to cope with these environmental salinity concentrations by having nature create a new cellular system. Such a cellular system would give aquatic birds an anatomical means by which they could survive in this new saline ecosystem, by secreting these new solutes from their vascular blood system.

In some unknown manner, nature through its mysterious adaptive genetic system has provided aquatic birds with a new type of epithelial cell that has been modified in its cytoplasmic composition, in which old subcellular organelles get new genetic messages that allow them to manufacture different types of protoplasmic pumps that can transport these newly ingested solutes from their bloodstream and empty them into the external environment. These basic differentiating cells, which have been triggered by unknown sensory cells, have become the adult extra-renal ionocytes.

This family of progenitors ionocytes have become the salt glands of the young chicks, and later evolved into the extra-renal organ of the adult bird. These salt glands, together with the adult kidney, give the aquatic bird the proper osmotic and ionic regulation of blood and lymph fluids in bird species living in hyper-saline wetlands (see scientific details in Chapter 4).

CHAPTER 2

Cell Structure of Ionocytes Found in Vertebrate Epithelium

by Frank P. Conte

Are the cellular structure and biochemical features of ionocytes found in all types of epithelial tissue similar in genetic response to sensory messages (miRNA), causing the cellular differentiation into the final adult secretory cell?

The Fish Gill Branchial Epithelial Ionocyte

The teleost fish gill has been studied in numerous species of bony fish. The genetic evidence begins with the fact that adult-type stem cell is the parent cell that becomes initiated as an ionocyte by cellular mitosis and continues development into various types of daughter cells that began the synthesis of ion-pump protein complexes. These protein complexes, which are intracellular manufactured, are then circulated to fit into the different locales of the outer cellular plasma membrane, to render the ion that is to be transported from the internal blood compartment and secreted into the external environment.

The discovery that a basal epithelial cell located at the base of the branchial epithelial cell mass was undergoing rapid mitotic replication in response to stress by external hyper-saline environment was made in 1967 by Conte and his coworkers, using an autoradiography technique with titrated H3-thymidine as a marker of cDNA replication.

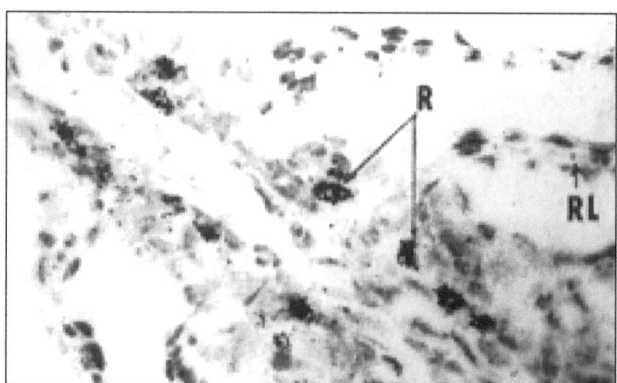

Figure 37. Replacement cell (R) with newly labeled cDNA in nucleus with H3-thymidine.

This established the site of the parental stem cell for the future adult ionocyte population. The F. P. Conte laboratory (Peterson, et al., 1978), together with H. E. Hokin's laboratory at the University of Wisconsin (D. Y. Sun, et. al., 1991, 1992), continued their work to show that the ion-pump complexes were found in these differentiating daughter epithelial cells, which contained radioactive-labeled protein ion-transporting complexes in epithelial cells, in both vertebrate and invertebrate types of salt glands.

Model of Basal Stem Cell of Ionocyte

The hypothetical model for the control of the growth and differentiation of the base stem cell of the ionocyte (or basic stem cell in chloride cell population in teleost gill tissue) was thought to be based upon a tissue-specific chalone system. The majority of epithelial populations measured under stable physiological conditions during 1965–1975 exhibited a cellular division rate of cDNA that is characterized as being very slow (T1/2 rate of days or weeks). But in unusual physiological situations, such as where cell damage or cell stress occurred, these populations underwent rapid cellular mitotic division (T1/2 rates of hours or single days). How could this difference in mitotic rate occur? It was thought that a diffusible substance (a chalone or pseudo-hormone-like substance) was released into the blood system, which could circulate among the cells of the body and modify the rate of cDNA mitosis. This modification could change the rate of ribonucleic messages or allow modification of protein biosyntheses. The exact sensory mechanism in the nervous system that provides the right type of molecular message was being delivered between cells in teleost gill tissue, and the saline environment was not known at the time. Therefore, the feedback circuitry between differentiated cells (or stem cells) in the developing gill and the final adult secretory cells was unclear (Lahlou, ed. Conte ch. 1980).

Differentiation of Progenitor Stem Cells of Fish Ionocytes

During the time period 1970–1990 there occurred a great deal of biochemical purification of ion-pump proteins and other membrane lipid-protein complexes that provided investigations as to the presence or existence of these basic stem-cell progenitors. A review of the origin and differentiation of ionocytes in gill epithelium of teleost fish was published by Conte in 2012. It provided the latest information on the biochemistry and molecular biostructures of epithelial cells undergoing salinity adaptation from the embryonic development of opercular/skin epithelium and the gill filamental/branchial epithelium. In some of the early investigations, the genomic pathways underlying the newly synthesized ion pumps were found to exist in different types of gill progenitor cells, indicating that different pathways must exist to create four differentiated stem-cell progenitors terminating in the adult ionocytes.

Four Types of Adult Fish Ionocytes

Lastly, P. P. Hwang and M. Y. Chou (2013) and P. P. Hwang, T. H. Lee, and L. Y. Lin (2011) have provided the most up-to-date information on these four types of adult ionocytes expressing distinct sets of transporters. They used zebra fish as their experimental model, to study integrative and regulatory physiology on the epithelial transport associated with body fluid homeostasis when zebra fish occupy freshwater and then become immersed in seawater during growth from embryonic stages to fully adult stages capable of living in both types of environments.

During embryonic development, ionocyte progenitors develop from epidermal stem cells and then differentiate into different types of ionocytes through a positive regulatory loop of FOS3b and other transcription factors. Several types of chalone and other hormones, including cortical, vitamin D, stannocalcin-1, and calcitonin, were found to participate in the control pathways of ionic homeostasis by precisely studying the target ion transport pathways, ion transporter, or ionocytes of the hormonal action. The topic of ion regulation in zebra fish has been the subject of numerous reviews, which are listed in this book. The four types of ionocytes found in zebra fish are analogous to other ionocytes listed for renal tubular cells that are found in various types of tissues in other vertebrate species.

The Bird Orbital Nasal Gland Epithelial Ionocytes

The scientific discoveries that revealed how the aquatic avian species with salt glands developed new cellular tissues, and which new epithelial cells were found in their salt glands, took place over four distinct time periods.

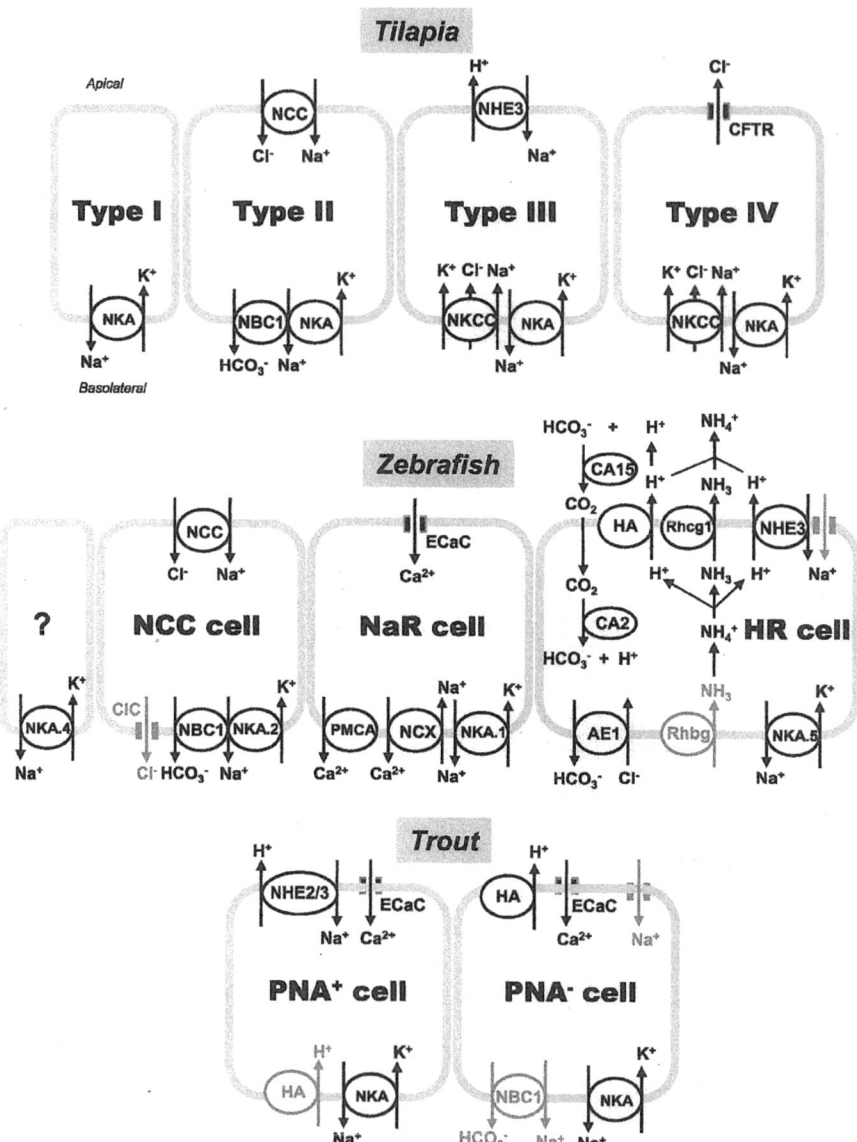

Figure 38. Models of ionocytes in various species of fish.

First, in the 1970s the location of new epithelial cells by thymidine -C14 cDNA stimulation under a new salt load was shown. Second, between 1978 and 1982 it was shown by autoradiography that incorporation took place in the lobular tubules, where the progenitor stem cells were identified. Third, in the late 1980s, the cellular modification of protoplasmic pump synthesis was identified, along with where it took place intracelluarly in the tubular

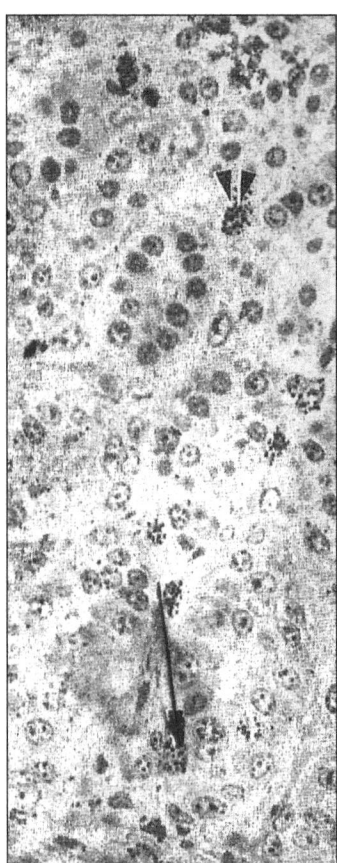

Figure 39. Autoradiography of H3-thymidine incorporation into lobular tubular cells of the ducklings.

cellPs. Fourth, in the late 1990s, the identification of the transgenic pathways that appear to be used by the adult ionocytes was shown. There appear to be three time periods in which the embryonic, ionic-pump protein transgenic pathways were discovered following the functional physiology that was uncovered by the Schmidt-Nielsen group at Duke University (1997). The question for the investigators of embryonic development of avian glands was that of cellular structure. Is it due to the hyperplasia within a single cell type, or is it mitotic cell division (cDNA) of a parental stem cell that made daughter progenitor cells develop into a series of pro-secretory cells and then finalize as the adult salt secretory cell?

Thymidine Tagging of cDNA in Avian Salt Gland

In the late 1970s, M. Peaker and coworker (C. Knight and M. Peaker, 1978) attacked the hyperplasia question through the use of H3-thymidine to label salt gland cDNA during the adaptation of adult ducks and geese that had been drinking either freshwater or saltwater prior to the injection of the compound. Different types of drinking schedules for the two environmental waters of FW or SW were used to challenge the incorporation pathways being marked by the radioactive label. They found that the salt glands had increased the cellular DNA by forty percent via the salt stimulation. The increase was due to cellular division (mitosis) and was not due to enlargement of a singular cell because of hyperplasia.

Auto-radiography of Avian Salt Gland Epithelium

Unfortunately, at this point, the types of labeled cells in the salt gland tissues were not followed by auto-radiographic analysis but were performed by F. Hossler's group at East Tennessee State (F. Hossler, M. Sarras, and F. Allen, 1978). Hossler and coworkers (1982) then pursued the question of ionic-pump protein synthesis and ribonucleic acid synthesis in the plasma membrane synthesis in the developing cells during salt stimulation. The measurements of protein were made by H3-leucine, and those of ribo-nucleic acid (miRNA) were made by triturated H3-uridine. The period of synthesis followed was twelve hours with labeled glycoproteins appearing after two hours, and this continued for a period of up to seven or nine days. Light auto-radiographic analysis of pulse-chase experiments showed that principal secretory cells were making new proteins and nucleic acids, and that glycoproteins were time-wise found sequentially, first being made in rER, then to cytoplasm surrounding Golgi apparatus, and lastly to the surface boundaries of the plasma membrane. At that time the group could not follow the biogenetic pathways that would give information as to the transgenic machinery being developed in these avian progenitor cells.

Avian Salt Tubular System Contains Ionocyte Biogenesis

J. Hildebrandt and coworkers (1998) picked up the problem of transgenic pathways in avian salt secretary cells when comparing the responses of avian salt glands and function to that which P. P. Hwang was publishing on the fish gill adult secretory cell was exhibiting in zebra fish embryos and adult fish. It appears that in the avian salt gland, the progenitor stem ionocyte showed research evidence that the cellular signaling machinery that mediated adaptive changes as observed in the biogenesis of the plasma membrane of the peripheral secretory cell found in the blind end of the tubule system is different than in zebra fish gills.

They found that the changes resulted in an enlargement and elongation of the salt secretory organ system, needed to be longer than fish gill since it took more time to increase the salt secretory capacity of the avian salt tubular system. Since the nasal orbital salt gland is enervated by parasym-pathetic nerves, it was necessary to determine how the gland regulated salt secretion. Was it parasympathetic nerves releasing acetylcholine transmit-ter? Or was it elaboration of environmental chalones (cues) from the brain acting on the plasma membrane surface sensory receptor sites?

These scientists investigated the ACh-mediated pathways by looking for muscarinic AC receptor activity in cultured nasal gland tissues. They found evidence that activation of c-fos genes had occurred, but Jun genes were not affected. It was concluded that Ach-signaling factors were playing an important role in the initiating development of cellular differentiation of avian stem cell in the salt gland tubular secretion.

Continued investigations by these workers showed that intracellular $Ca++$ ion increases upon ACh stimulation of muscarinic receptors had occurred and that the $Ca++$ ion entry mechanism was derived from in-tracellular stores and its involvement with inositol phosphate molecules. This mechanism must play an important role in the adaptive process of making stem cell progenitors needed in the salt secretion pathway for the formation of adult stem cell as the final salt-secreting ionocyte. Recently, the Hildebrandt group of investigators have shown that the FOS expression is pronounced in the formation of the adult stem cell ionocytes in the nasal gland epithelium, and that other cell types—in particular, vascular endothelial cells—show FOS expression as well.

This demonstrates that virtually all cell types in avian glands' cellular differentiation are involved in the process of salt secretion and its delivery to the excretory duct system of the nasal beak system. Continued development of the nanoscopic microscopy of cellular functionality will give us more elucidation as to intracellular events at the molecular and genetic level.

CHAPTER 3

Migratory Birds and Energy Required for Sustaining Life Cycle

by Frank P. Conte and A. Henry

Thermal Energy

The various climate changes in the environmental geological regions for birds, especially aquatic habitats in glacial areas, have made for tremendous variations in heat conditions. In summer months these areas will reach 100°C, while in the winter season temperatures will drop below freezing (0°C). Thus, nature required birds to seek other seasonal temperatures for winter sites. Doing so saves the thermal energy used in holding high body temperatures. In this case, by maintaining constant body temperatures, which uses thermal energy, it can transfer the energy into aerial flights for locating warmer regions that usually abound in the southern tropical or semi-tropical aquatic habitats. Most aquatic avian species have shown in ornithological investigations that this physiological adaptation has occurred (M. Ramenofsky and J. C. Wingfield, 2007).

Winter and Summer Grounds at Lake Abert for Migratory Species
(Boula and Jarvis, 1984 and R. L. Pers. Comm.)
The number of migratory birds that have visited Abert Lake has been monitored by ornithologists who have visited the lake and are cited in the tables below.

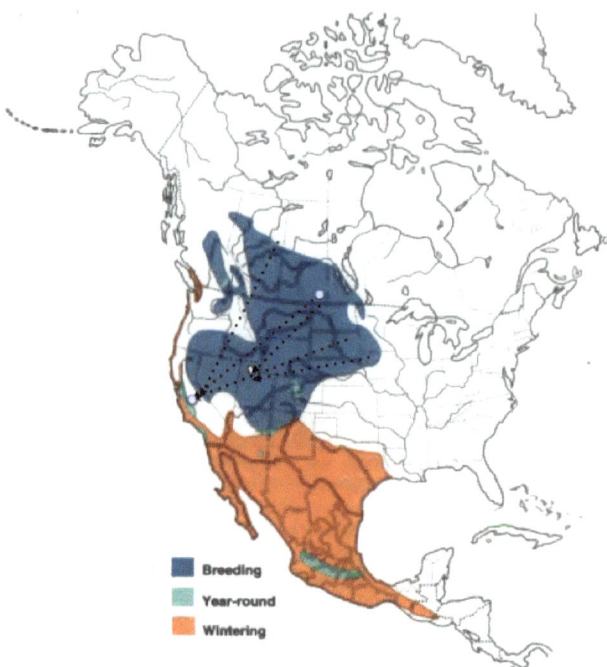

Figure 40. Breeding, year-round and wintering seasons. (Courtesy of A. Henry)

Winter and Summer Grounds for Grebes at Lake Abert and Mono Lake

Figure 41. Adult grebe nest and egg. (Courtesy of A. Henry)

Figure 42. Adult grebe with chicks, searching for food. (Courtesy of A. Henry)

Figure 43. Grebes searching for food sources—brine shrimp and brine flies. (Courtesy of A. Henry)

Figure 44. Eared grebe—developed foraging strategy of diving for brine shrimp. (Courtesy of A. Henry)

Lake Abert serves the important function of allowing migrating birds to optimize their condition before continuing their long-distance migrations (Jehl 1988, pp. 52–53). This is especially important for small, shorebirds whose long-distance migration may range from the Arctic to South America. Boula and Jarvis (1984) found that the fall diet of most migrating birds at Lake Abert was dominated by the alkali fly. Brine shrimp occurred

Species	No. of birds[1]	Date	Source
American avocet *(Recurvirostra americana)*	15,345	Aug 97	Warnock et al. 1998
western sandpiper *(Calidris mauri)*	15,000[2]	May 89	Kristensen et al. 1991
Wilson's phalarope *(Phalaropus tricolor)*	60,000-70,000	Jul 82	Jehl 1988
red-necked phalarope *(Phalaropus lobatus)*	43,450	Aug 95	Sullivan 1996
eared grebe *(Podiceps nigricollis)*	30,000	Apr 94	Tweit et al. 1994
ring-billed gull *(Larus delawarensis)*	20,000	Fall 96	Sullivan 1996
California gull *(Larus californicus)*	8,000	NA[3]	Marshall et al. 2006

[1] Single-day counts

[2] Count consists "mostly" of western sandpipers

[3] Data not available

Table 3. Some Reported Maximum Numbers of Waterbirds Observed at Lake Abert. (Pers. Comm., R. L.)

frequently only in the diets of northern shovelers and eared grebes (G. Keister, 1992).

Following are several summer photos of avocets at Lake Abert (photos from OPB video and R. L., Pers. Comm.).

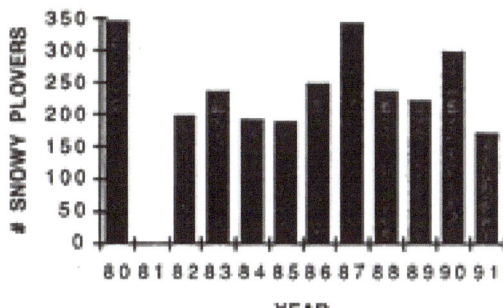

Figure 45. Snowy plovers counted at Lake Abert, Oregon, in June 1980–1991. (Data for 1980 from Herman et al., from ODFW counts)

Power Energy

Flight patterns for any distance require the development of anatomical flight muscles that support body weight. The other physiological functional needs of the bird's body, such as osmotic and ionic systems of salt glands to provide the necessary body water and fuel metabolites for flight, must also be sustained.

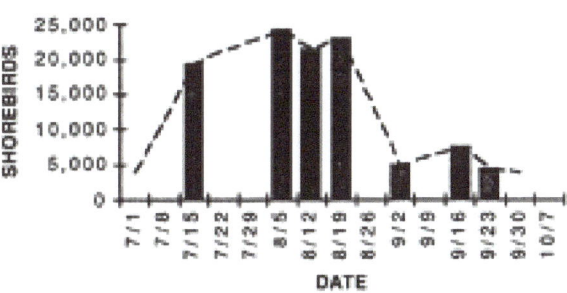

Figure 46. Total number of shorebirds at Lake Abert, Oregon, 1991. (Data from Morawski and Stern, 1991)

Figure 47. Summer photo of avocet on Lake Abert. (R. L., Pers. Comm.)

Figure 48. Summer photo of avocets and shovelers flying at Lake Abert. (Courtesy of Ron Larson)

Figure 49. Summer photo of avocet nest at Lake Abert. (Courtesy of Ron Larson)

Figure 50. Avocet on nest at Lake Abert. (Courtesy of Ron Larson)

Figure 51. Flocks of avocets and phalaropes feeding at Lake Abert. (Courtesy of Ron Larson)

Figure 52. Phalaropes resting on Lake Abert rock pile. (Courtesy of Ron Larson)

Figure 53. Single phalarope feathering at Lake Abert. (Courtesy of Ron Larson)

Figure 54. Phalarope resting on water at Lake Abert. (Courtesy of Ron Larson)

Figure 55. Flock of grebes resting at Lake Abert. (Courtesy of Ron Larson)

Figure 56. Flying golden eagle. (Courtesy of Ron Larson)

Anatomical Flight Muscles

The pattern of muscles for avian flight has been shown in studies by A. Biewener (2011, vol. 366:1570, published 18 April 2011. DOI: 10.1098/rstb.2010.0353), revealing that the flapping anatomical muscles of the wing system place strenuous requirements on the animal's energy system. In birds, this energy demand is greater than the leg muscles needed to walk or jump over and around terrestrial rocks. Thus, when contrasted with a terrestrial locomotion system, which offers many mechanisms to reduce the energy requirements, such as saving elastic energy used in lifting the skeletal body units, the flight aerial system must make use of this extra energy. In order for birds to produce the aerodynamic power needed to support their weight in the air and to overcome drag, they need a special kind of mechanical system. That system needs to be one that will generally contract at high frequencies and be independent of organ size.

The flight muscles in birds use the large pectoralis muscles, located in the shoulder, to depress the wings. The dominant role and large size of the pectoralis muscle, therefore, enables one to establish an investigative fiber-system. Since pectoralis is a large muscle (10+ percent of body size) and is attached to the humerus of the wing at the deltopectoral crest (DPC) and with other anatomical appendages that permeate the other sides of the wing, Biewener and his associates were able to attach an analytical strain-force instrument to measure the recordings of the pectoralis and the supra-coracoideus. Using sonomicrometry sensory transducers, they were able to measure fascile strain rather than muscle fiber strain.

This group of scientists continued working the instrumentation such that, for a number of bird species and different body sizes, they found for those species studied, the *in vivo* force-strength work behavior of the pec-toralis is generally similar across a range of flight speeds and conditions. In summary, these scientists believe that the aerodynamic and metabolic power requirements for flight are of considerable interest to avian and evolutionary ecologists interested in the strategies that birds use to forage for food and migrate to ensure a successful life history.

Chemical Energy: Energy in Flight and Energy of Body Functions in Food Sources

Flight patterns are dependent upon generating the needed goals for the life of the bird in flight. However, the flight pattern must survive duration of environmental climate conditions while in flight. What happens after flight? What is the type of energy that supports the lifecycle requirements, such as reproduction of the adult, egg and chick development, and adult defensive maneuvers for avoiding death by predators. The major chemi-

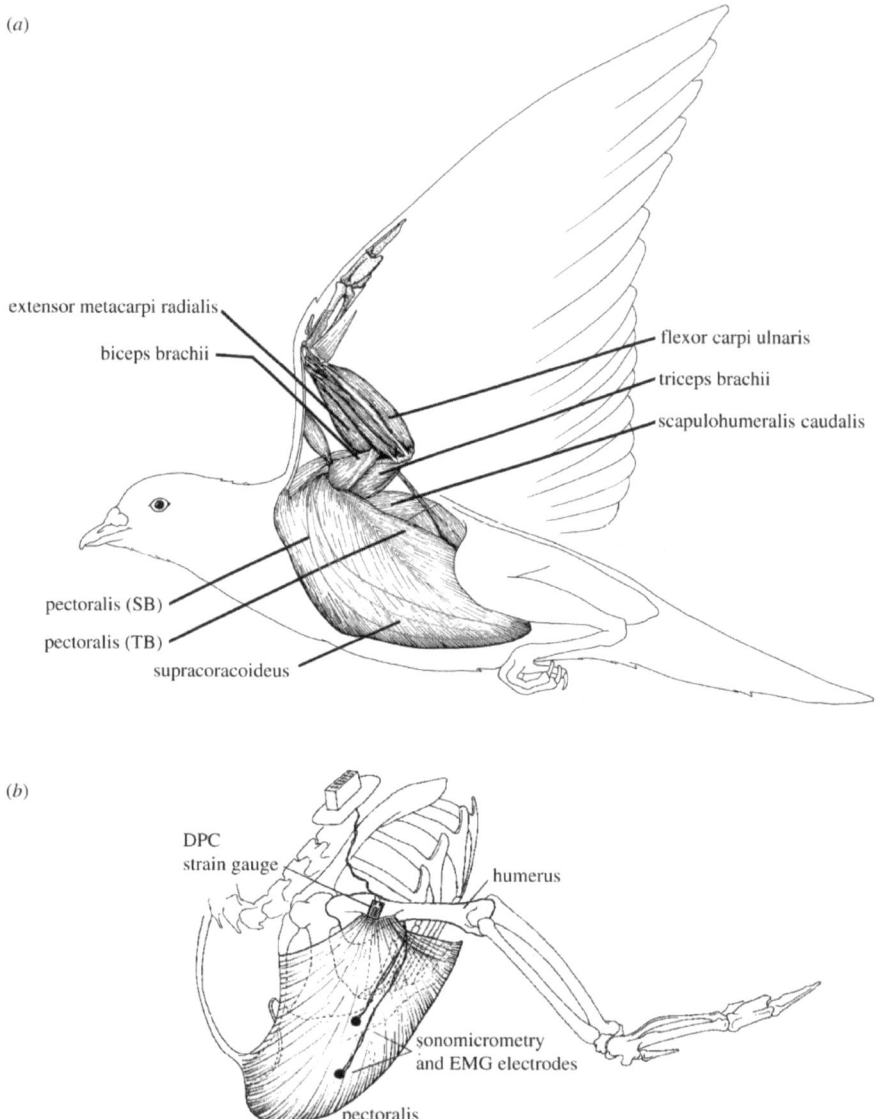

Figure 57. Biewener shows the anatomical location of muscle in relation to

cal energy required for these conditions comes from the digestion of food captured by the adult and the metabolic pathways of digestion available to the individual bird weathering these conditions.

1. Freshwater comes from the renal and extra-renal organs, which are the kidney and the salt gland. Both of these organs are needed to maintain osmotic and ionic regulation of the body fluids. They utilize the metabolic

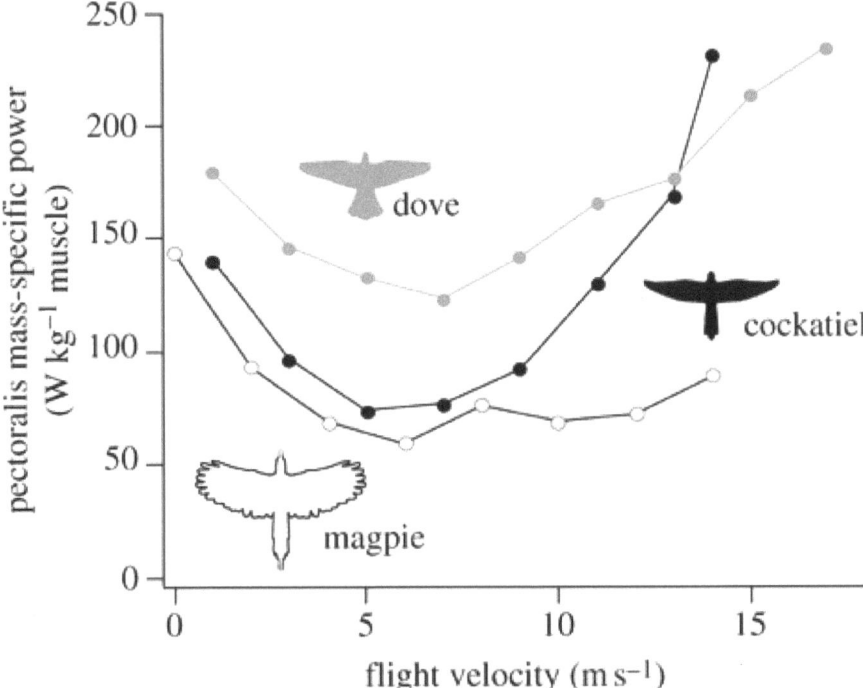

Figure 58. Flight velocity of birds.

energy coming from the mitochondria, ATP. The metabolic food source is carbohydrates, fats, and proteins. In addition, oxygen is needed.

2. Egg development and embryonic growth are biological processes that need water and basic chemical products from the digestive food tract for the continuance of the avian adult life cycle. Avian food sources have been published by ornithologists interested in the biology of adult birds and are listed in several of the book selections identified in the references. The reader may wish to refer to these publications for more details on avian food and metabolism.

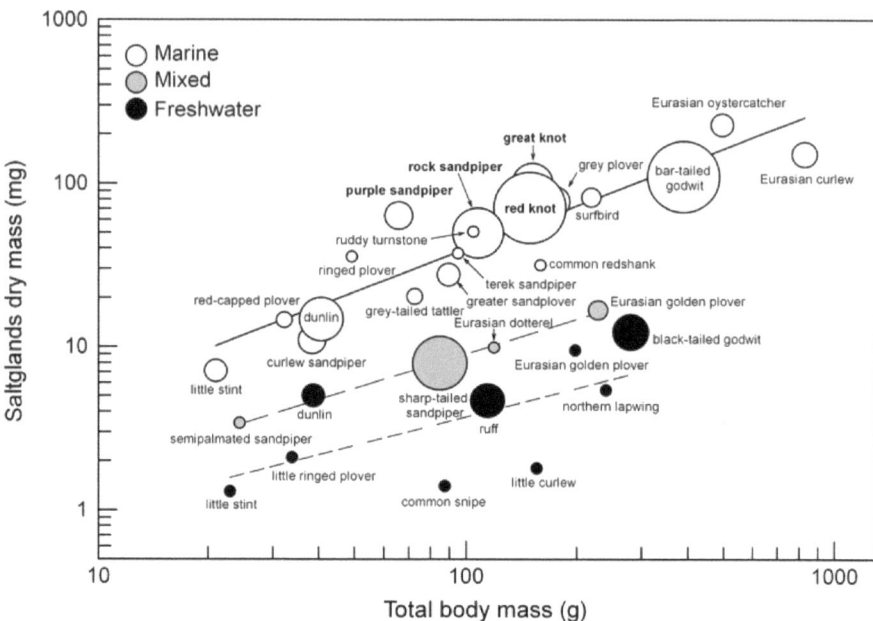

Figure 59. The allometric relationship between log dry msg and log mb for 29 shorebird species obtained with conventional regressions. Regression lines for marine species, log dry msg = -0.352 + 0.876 log mb (r2 = 0.816, P < 0.0001; solid line); 'mixed' species, log dry msg = -1.046 + 0.705 log mb (r2 = 0.997, P < 0.001; long-dashed line); and freshwater species, log dry msg = 0.412 + 0.412 log mb (r2 = 0.074, P = 0.241; short-dashed line). Results were similar using phylogenetic contrasts (see Table 1). Symbols are scaled relative to the sample size, except for bar-tailed godwit (N = 133) and red knot (N = 185). Marine molluscivores that typically ingest prey whole and those eating soft-bodied invertebrates are depicted in bold and regular characters, respectively. Note that values for the little stint, the dunlin and the Eurasian golden plover are represented for more than one habitat type and are subsequently treated as separate in the analyses.

PART III
Halobionts Inhabiting Salt Lakes: Food Sources of Avian Migratory Birds

CHAPTER 1

Halobionts:
Life or Death due to Salinity

by Frank P. Conte

A salt lake is a landlocked bowl of water that has concentration of salts, typically sodium chloride, and other dissolved minerals significantly higher than most dissolved minerals found in seawater (32.5g per liter). Any lake that has more salts than seawater is termed *hypersaline*. Most salt lakes have some measure of freshwater flowing into them. If the amount of water flowing into them is less than the amount being evaporated, and if the water left behind cannot leave because the lake is dead-end or terminal (endorheic), the solutes remaining increase in salinity.

High salinity will also lead to unique halophilic bacteria, flora, and fauna in the lake in question. But if the freshwater entering the lake is reduced to an amount less than that which is evaporated, the lake will become desiccated to the point of complete dryness, in which case the lake bed is called a *salt flat* or a *playa*.

Halobacteria
Then, all halophilic flora and animal fauna die and only microscopic bacteria and algal forms and other archaic forms will continue to exist. Lake Abert in southern Oregon is undergoing desiccation, as is shown in the photographs taken this past year.

Figure 60. Abert Lake, drying up due to climate conditions (photo taken July 30, 2014. The red playa is caused by halobacteria (species not identified). (R. L., pers. comm.)

This picture shows the most extreme of halophiles, which are often called halobacteria, and can be seen as the red mass that is lying upon the Lake Playa. It is a group that is scientifically referred to as Archaea, which are the oldest type of halophiles and usually require at least a 2 M salt concentration in which to live. They are the primary inhabitants of salt lakes and inland seas. These prokaryotes require salt for growth. They also equire oxygen for growth. Their cellular structure has internal machinery that is genetically adapted to having charged amino acids (red color), with electric charges attached to the lipid surface of their external plasma membranes so that water molecules can stay attached around these components. Unfortunately, these organisms are toxic to many other types of plants and animals and cannot be used as food sources.

Halophilic Algae

The major plant source that survives in these very high-alkaline salinities are the various species of chlorophyres and cyanophyts (Herbst, 1986). In Lake Abert, it was found that the major algal specie was the chlorophyte, which is referred to as the filamentous green algae *Ctenocladus circinnatus*. It is mostly found in the graveled shoreline type of mudflats, attached to benthic rocks lying beneath the water surface, or in rocky regions of bottom salt beds. The appearance of this filamentous green algae is that of a green mat or rug just covering the bottom of the lake or the entire rock surface. Vegetable cell growth of *Ctenocladus* occurs in the form of branching filaments, which are short segments of thick-walled terminal resting

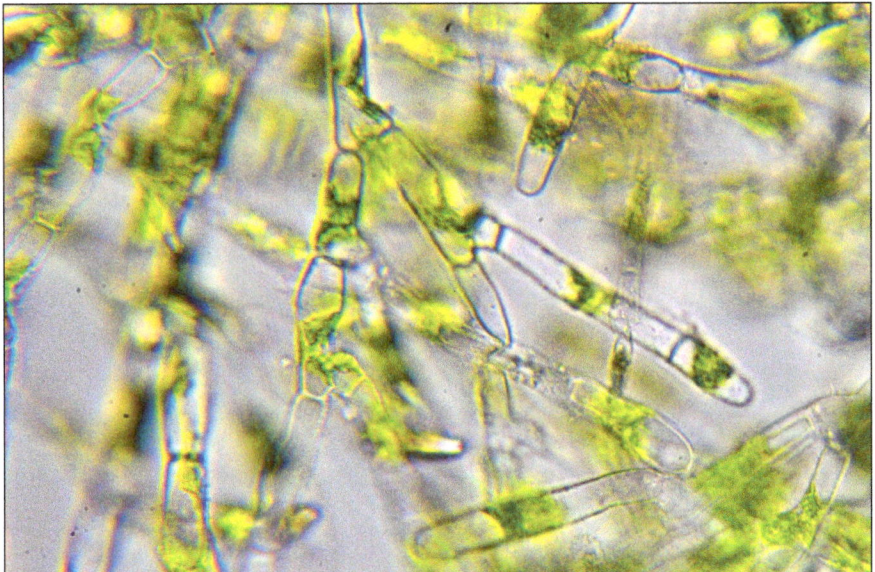

Figure 61. *Ctenocladus* algae. (Courtesy of Keith Kreuz)

akinetic cell types, often in chain-like rows as if they were branches of a tree growing underwater. The dimensions of these cells may vary between locations, but it is mostly environmental conditions that produce the energetic sub-states required for these growing cells. These akinetes act as diapause embryos since they can resist damage from salt-saturated brines, complete desiccation, and freezing. Germination of the akinetes can occur upon rehydration accompanying warmer temperatures. Various scientific investigations have shown different environmental factors can influence the breakage of diapause and growth of these akinetic cells (Blinn, 1971). The cells of this green algae are the biological surface structure that is most important for the development of the brine fly, and for brine shrimp populations to interact with in starting their unique life cycles.

Halophilic Brine Fly (*Ephydra*)—One of the Top Avian Food Sources

The physiological ecology identifies the internal and external limitations on the acquisition of resources needed by the inhabitants for use as food. As for migratory birds, the brine fly is one of their major food sources. The flies use the green mat of algae as their major physiological ecological region for growth, reproduction, and continued life cycle, beginning in the spring and lasting until the cold winter months. Thermal regulation is a major factor in the growth, development, and size of adult flies from the larval and pupal type of juvenile stages. Therefore, sunshine and temperature are important environmental factors. Brine fly use osmotic and ionic regulation of external water and solute concentrations and require that the external environment continue to contain water. Even if the lake is drying up due to drought climatic conditions, there must be water for brine fly stages to continue to sustain internal osmotic and ionic regulation of blood throughout the growing season. If external alkaline salinity exceeds 150 g/liter, the survival of larvae is substantially reduced and adult stages, if they survive, have a much reduced size. If birds eat the larvae, their energy intake is greatly reduced for energetic demand. The reduction in food energy has a marked impact upon migratory flight distances. Herbst (1976) has found that Abert Lake larvae are less tolerant of increased salinity than larvae taken from other, comparable saline lakes (e.g., Mono Lake). But the pupal and adult fly size were larger at Lake Abert, as were the regions for fat storage.

It was interpreted that the increase in these food storages must be due to the high benthic algal food resources the various stages of brine fly adult stages had been using.

Insect Lime Gland: A New Extra-Renal Organ for Salt Secretions

The structures responsible for osmotic and ionic regulation of the internal fluids have been investigated by Herbst and Bradley (1989). They showed that brine fly do not have a salt gland organ. They have an extra-renal organ called a lime gland. They have genetic messages that control the development of the intestinal tract to enable tubular cells to capture hydroxyl ions.

Larva

Adult

Figure 62. *Ephydrahians*. (Courtesy of Herbst and Bradley)

These investigators showed that within the malpighian section of the ureter tract, there were cells that captured alkali-producing hydroxyl ion with metabolic carbon dioxide to form carbonate ion. This permits the conversion of carbonate ions to bind with calcium ions to form the insoluble salt that can be deposited as lime crystals in the anterior region

Figure 63. Adult brine fly feeding on green mat of *Ctenaclatus*.

of the renal tubular system, which also secretes soluble sodium chloride. These reactions become the means to control osmotic and ionic regulation of blood and lymph fluids.

Thus, the eggs deposited by adult brine fly can pupate on the green mats of *Ctenocladus* and continue larval development into future adult flies. These adult flies are the source of food that the migratory birds feed on during late summer months. They contain large amours of fats and proteins.

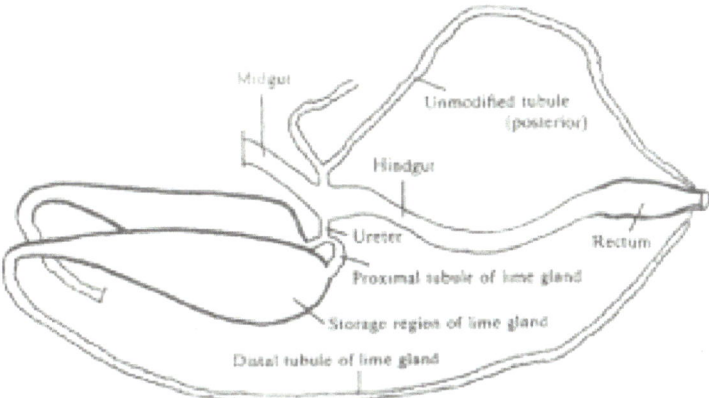

Figure 64. Schematic diagram of malpighian tubules of *Ehydra hians*, showing both the modified pair of lime gland tubules (anterior) and the normal pair of unmodified tubules (posterior).

Halophilic, Branchiopod Crustaceans—Daphnia, Fairy Shrimp, and Brine Shrimp

There are several species of branchiopod crustaceans, known as shrimp, that have similar limitations in their tolerance of lake water salinity. The species that are least tolerant of salt and live in Lake Abert when at its highest water level are the daphnia and fairy shrimp (see Figure 65).

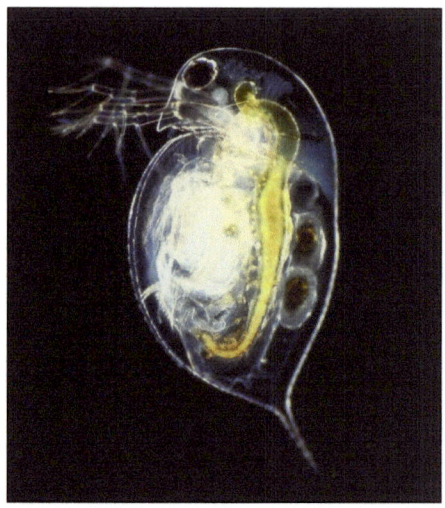

Figure 65. Daphnia (courtesy of Ron Larson and Frank P. Conte)

CHAPTER 2

Brine Shrimp Population in Lake Abert: Top Food Source for Migratory Avian Species

By Frank P. Conte

The shrimp population that exists in Lake Abert in the greatest abundance is the branchiopod crustacean known as the brine shrimp (*Artemia salina*). This population of these shrimp in Lake Abert is as important a food source as the brine fly. The brine shrimp population became an important economic commodity for the State of Oregon when it was shown by Oregon Sea Grant investigators (F. P. Conte and colleagues, 1978–1982) that, at a salinity ranging from 40 g/L to 80 g/L, Lake Abert's brine shrimp population had a calculated mass of approximately 14 million pounds.

These estimates, together with the data Keith Kreuz and associates collected at the lake during the same period of time, provided the impetus for establishing the Oregon Desert Brine Shrimp Company (see Part IV), the first inland bait fishery for brine shrimp.

Figure 66. Fairy shrimp. (Courtesy of Ron Larson and Frank P. Conte)

Figure 67. Male (left) and female (right) brine shrimp. (Courtesy of Keith Kreuz and Ron Larson)

Figure 68. OSU Sea Grant sign for Brine Shrimp Project and Cooperators.

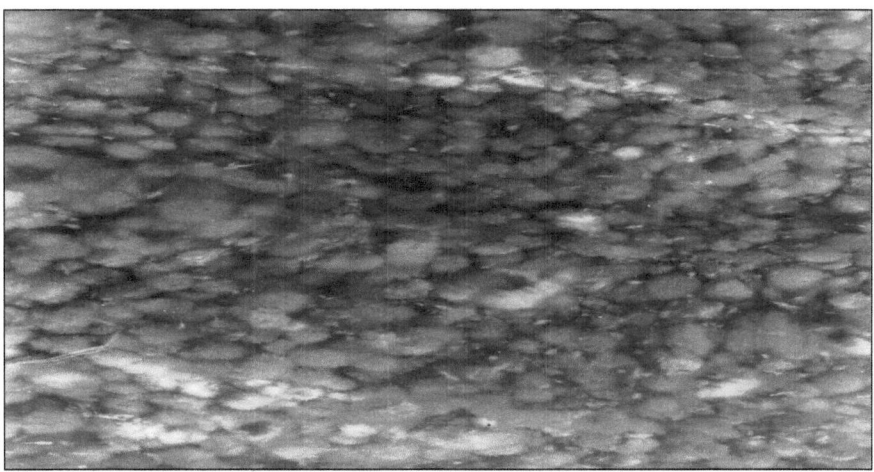

Figure 69. Lake Abert playa showing brine shrimp feeding on floored algal blooms.

Figure 70. Frank P. Conte measures the salinity of Lake Abert.

PART IV
Lower Chewaucan River Freshwater Management Issues

CHAPTER 1

Changes in Federal and State Legislation Used by Management Agencies

by Frank P. Conte

Environmental Changes Requested by Federal Agencies

Several legislative management changes have been requested by two federal agencies, the Environmental Protection Agency and the US Army Corps of Engineers. They are responsible, under the Freshwater Act of 1972, Section 400, for the management of the nation's watershed activity. The details of the cited document, "Connectivity of Stream and Wetlands to Downstream Waters: A Review and Synthesis of the Scientific Evidence" (2013), were derived from a large number of publications that had been peer-reviewed by both scientists and literature professionals. Most of the literature covered in this document emphasized the physical and chemical characteristics and physical functioning of downstream waters. The document was to serve as a means of awakening the public to some of the difficulties the agencies had in certain watershed activities. For instant, some of these waters that included perennial, intermittent, and ephemeral lakes did not have clarity as to their legislative duties. The term *connectivity* is a fundamental concept in hydrology and freshwater ecology because it is believed that the structure and function of downstream waters are highly dependent on the constituent materials contributed by and transported through water bodies located elsewhere in the watershed. An example of this concept is an ephemeral river basin such as Arizona's San Pedro River Basin. It was cited as an ephemeral river basin that had been under management focus for a long period of time. However, similar regions in other Western states have had long periods of scientific investigations and were not cited in this report. We believe that the Lower Chewaucan River Basin, in south-central Oregon, has had a longer period of scientific investigation. It should have been given equal consideration in support of these conceptual legislative changes in the CWA 1972-Section 400. Lake Abert is an ephemeral lake whose freshwater inflow comes from the Lower Chewaucan River. This river basin has been studied since the early 1900s, as is described in Chapter 1 of this book. Therefore, the High Lakes Aquatic and Alliance Foundation (HLAAF) and its non-profit volunteer group voted to have the organization sponsor this report. It has submitted it to the Environmental Agency (EPA) with an elaborate summary that deals with the hydrology and ecology of this basin during the past hundred years. It was accepted by the agency after receiving our document as a scientific report as a WOTUS. EPA change of rules in CWA, 1972-ID, EPA-HQ-OW-2011-) 880, dated November 7, 2014.

Live Bait Industry—Oregon Desert Brine Shrimp Co.
by Keith Kreuz

History of the Formation of the Company
In 1979, Keith Kreuz and a single associate began researching the waters of Lake Abert for the presence, numbers, and sizes of adult brine shrimp. In 1980, they informed Oregon Sea Grant of their intent to establish a private company to harvest brine shrimp in Lake Abert. The purpose of the letter was to inform the director of Oregon Sea Grant that the awarding of the brine shrimp research project to Professor Frank P. Conte might harm their potential of establishing a bait fishery company. Dr. Conte replied in a letter date September 1980 that his research project would establish that the population measurements of brine shrimp his group found would not hinder the formation of his company but would either contribute data in support of his effort or show that the lake's water could not accommodate his claim. Thus, a collaboration between the two groups began, and subsequently the publication of an article in *Hydrobiologia* (1982) by F. P. Conte and P. A. Conte on the abundance and spatial distribution of brine shrimp (*Artemia*) in Lake Abert was used to support Kreuz's bid to establish the Oregon Desert Brine Shrimp Company in 1979–80. The company ultimately became a commercially successful bait fishery, until Lake Abert dried up in 2014.

Development of a Brine Shrimp Trawling Harvester

Early in 1979–80, Keith Kreuz designed and built a motorized trawler for catching small adult shrimp (5–10 microns). The trawler was designed to withstand the weather and salinity and alkalinity of Lake Abert's waters and also catch large numbers of shrimp. The craft Kreuz designed and used was unique in its structural simplicity and low cost. It consisted of a plywood platform that could haul appropriate trawling nets with cod-catching ends and empty them into holding tanks on the deck of the fishing platform, which also held the cockpit for pilot-crew and a motor for moving through the lake's water. He started out with a plywood raft platform of suitable dimensions to hold up to twelve five-gallon plastic buckets, which would be filled with shrimp and then transported first to the dock and then to a nearby packing plant. The plywood platform was held together with four-by-six styrofoam logs for flotation. Ten to twelve empty five-gallon plastic buckets would line the deck of the platform, and in the center of the raft was the cockpit for the pilot-crew member to operate the eight-horsepower outboard motor for boat mobility. Thus, the pilot-operator could move the raft to various locations on the lake's surface where the brine shrimp would be feeding.

To capture shrimp, the twenty-foot trawling net had openings large enough to catch adult shrimp of eight to ten millimeters in size. It also had a cod-end plug, with four dowels and two nets attached to frames built from an eight-foot length and two-foot width of stainless steel, supported by the dowels on each side of the raft. Thus, the nets could extend the full length of twenty feet behind the raft and then be hauled into the raft for emptying shrimp into the buckets. The following photographs show

Figure 71. Photo showing Kreuz operating shrimp trawler over fishing area.

Figure 72. Keith Kreuz pilots his shrimp trawler on Lake Abert.

Figure 73. Keith Kreuz emptying trawl nets into five-gallon buckets.

Keith Kreuz operated the trawler platform in this text, but the reader can enjoy a video produced by Oregon Public Broadcasting's *Oregon Field Guide* in 1987 (Episode 405). The operation of a brine shrimp fishery is depicted in full color and with Keith's own commentary on how this business operates as he talks with Steve Amen, the host of *Oregon Field Guide* (K. K., Pers. Comm. 2015).

Figure 74. Harvested brine shrimp on their way to a processing plant in Valley Falls, Oregon.

Harvest Season for Brine Shrimp

Lake Abert's brine shrimp populations, and the amount of shrimp that can be harvested, are not directly correlated with the salinity of the lake's water. We have found that brine shrimp catch rates and harvest quantity may be exceptionally good or poor, regardless of whether lake levels are high or low. As an example, during the high water years of 2000 to 2003, we experienced a record low harvest in 2001 and relatively high harvest yields in 2000 and 2002, despite water levels being approximately the same all three years. Exactly what environmental factors led to a reduction in brine shrimp populations in 2001, we still do not know.

Month of Year Brine Shrimp Are Born

The time of year that brine shrimp are born depends upon the anatomy and physiology of egg development in the female shrimp. For instance, when eggs are formed by the female early in the season and deposited in

Figure 75. Lynn Kreuz uses a Simplex packaging machine to place freshly caught and washed brine shrimp, which have been weighed, into plastic bags for freezing and storage at Valley Falls, Oregon.

her egg sac, they can be fertilized by the male and start embryogenesis. When an egg reaches the pre-naupliar stage, it is discharged into the water as a swimming nauplius, whereupon it continues its developmental growth until adulthood. However, if the female continues to make eggs during the approach of fall, when water temperature begin to drop, the fertilized eggs are kept in her egg sac and form gastrula-stage embryos. This stage contains an embryonic protein-lipid membrane, which surrounds the embryo and prevents further development. Thus the eggs are transformed into diapause embryos, which are dormant. When the female dies during the winter season, these trapped, shelled embryos (or cysts, also popularly known as "sea-monkeys") are released into the saline water, and the winter winds and water currents deliver them to the shoreline, where they become dried shrimp embryos. These shelled embryos continue to live but are acted upon by physical and chemical factors of the environment until suitable conditions bring about an "activated" embryo. These dried embryos have been studied by A. Warner, T. MacRae, and J. Baghaw (1989) (J. S. Clegg, 2011) and a group of scientists in Belgium, under the leadership of Patrick Sorgeloos, University of Ghent.

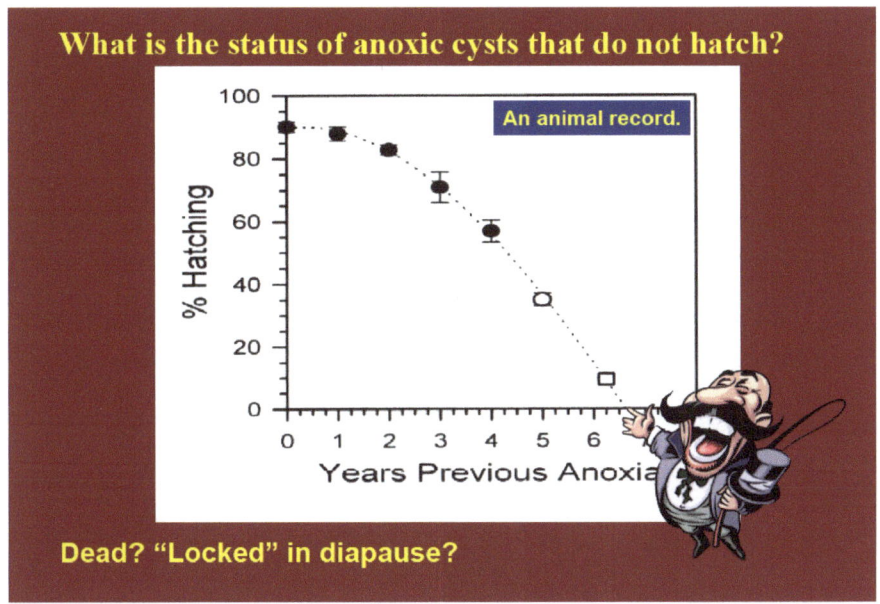

What is the status of anoxic cysts that do not hatch?

An animal record.

Dead? "Locked" in diapause?

Figure 76. This is a summary of the aging of dried shrimp and their hatchability for making a new population of shrimp. (Courtesy of J. Clegg, et al.)

The Months of the Year in which Adult Shrimp Are Harvestable

The month of the year in which an adult brine shrimp reaches the harvestable size of eight to ten millimeters in size varies from year to year. For instance, in 1992 the harvesting of brine shrimp from Lake Abert began in April, while in 2001 the harvest was delayed until mid-June. Similarly, a reduction (crash) in brine shrimp populations can occur as early as August 1 or extend into September. Again, we do not know what environmental factors affect population levels. Is it salinity, alkalinity, or temperature?

Harvesting generally begins before sunrise. Why? Possibly because the shrimp may favor a cooler temperature, or maybe it's to avoid the floating algae (which can also contaminate our catching nets). In addition, at sunrise the algae begins to produce oxygen in its metabolism, which lifts it off the lake's playa soil and rocks and makes it rise to the surface for more sunlight, aiding its photosynthesis.

This algal flotation phenomenon does affect harvesting. The amount of shrimp that can be harvested following a two-hour trawl is usually 400 to 500 pounds, all of which is stored in the buckets aboard the raft-trawler. After the raft-trawler has unloaded the shrimp into a truck for hauling to a nearby packing plant, the shrimp are then rinsed with freshwater to remove surface salinity and alkalinity and any other contaminants. Following drainage of the freshwater, the fresh adult shrimp are packaged in resealable bags, as shown in Figure 75. The shrimp are then repackaged in bags of various sizes, and are then frozen and stored in a factory freezer for shipment to Portland, Oregon, where they are shipped to customers.

Market for Frozen Oregon Desert Shrimp

Frozen brine shrimp are primarily sold to the tropical fish industry as a food product. Frozen shrimp are also used as a food source at federal and state fish hatcheries. Flake fish-food manufacturers also use frozen brine shrimp from Lake Abert in their products. A large volume of Lake Abert brine shrimp has been sold to live prawn producers throughout the world. Brine shrimp has proven to be an excellent food source in the raising of prawns for human consumption. However, some countries with saline lakes that produce and market brine shrimp have a reputation for some of their shrimp carrying disease organisms, which can contaminate domestic prawns and make consumers sick. Therefore, prawn producers are very interested in brine shrimp from Lake Abert, which have been tested to be disease- and contamination-free. Thus, Lake Abert has the potential to be a major supplier of disease-free, high-quality brine shrimp.

This bait fishery could also be a low-impact sustainable fishery. The biomass of brine shrimp in Lake Abert has been estimated at 14.5 million pounds (Conte and Conte, 1982), which Kreuz believes to be unreasonably low. His findings by ODB investigations during thirty-five years of commercial fishing on Lake Abert indicate that brine shrimp populations can vary from 10 million pounds to well over 100 million pounds per year. He estimates that Oregon Desert Brine Shrimp Company's total annual commercial harvest amounted to just a fraction of that—about 50,000 pounds.

Future Commercial Harvesting of Brine Shrimp

The future of brine shrimp harvesting from Lake Abert is rather on the bleak side. Since 2014, the water level in the lake has become too low and hypersaline (>100mg/liter), making it toxic to the shrimp. The Lower Chewaucan basin is experiencing climatic drought conditions, and freshwater inflow into the lake has diminished to virtually zero. Agriculturists and foresters with water-diversion permits have needs that must be met. What water resource managers need to determine is whether the freshwater flowing through Paisley, Oregon, into the Lower Chewaucan River is sufficient in volume based on existing and projected atmospheric and subterranean sources. Until this happens, it is unlikely Lake Abert will ever realize its potential as a major supplier of brine shrimp.

Figure 77. Agricultural areas along the Lower Chewaucan River, between Paisley and Valley Falls. (Courtesy of Dr. William A. Bowers, 2005)

Agricultural Lands: Farms and Ranches in the Lower Chewaucan River Basin

Following are three articles written by Lee Juillerat and published by the *Herald and News* on September 10, 11, and 12, 2015.

Abert: A lake gone dry

PART 1 — Alliance, ranchers dispute why Lake Abert is dry (Thursday, September 10, 2015)

LAKE ABERT — It's not a typical lake. Some years Lake Abert is an alluringly beautiful blue expanse of water, one that visibly shimmers on breezy afternoons.

But even during the years when it's partially or completely full, it's not a place to swim, boat or fish. Abert is a terminal lake, which means it has no outlet. A remnant of Lake Chewaucan, which 13,000 years ago covered an estimated 500 square miles and reached a maximum depth of 375 feet, Abert is much smaller, but still expansive. At its maximum recorded elevation in 1958, the lake covered 64 square miles and had a maximum depth of 16 feet.

Figure 78. Photo by Lee Juillerat. Ron Larson, with HLAAF, walks along the dry lake bed of Lake Abert in Lake County in August [2015]. The HLAAF suspects Paisley area ranches are using water at levels that threaten the lake's future.

Figure 79. Photo by Lee Juillerat. The Lower Chewaucan River flows from the River's End Ranch reservoir to Lake Abert in late May [2015]. Until the lower Chewaucan was settled in the 1870s, the upper river flowed into a large marsh that extended toward Lake Abert.

Figure 80. Photo by Lee Juillerat. The waters of Lake Abert are highly alkaline, too salty to support fish. Collections of salt crystals form as water evaporates from the lake surface.

Now, as then, its waters are highly alkaline, too salty to support fish. Eleven years before John C. Fremont's visit, John Work, leader of a Hudson's Bay Company fur trapping expedition, in his Oct. 1, 1832 journal dubbed it Salt Lake.

Abert's high salinity and alkali content isn't necessarily a detriment. Those conditions make it ideal for brine shrimp and shore flies, favorite food for migrant birds, including Wilson's and red-necked phalaropes, American avocets, killdeer, Ross' geese and eared grebes.

Last year, Lake Abert was parched dry. Thanks to spring rains, this past summer the lake had some content, but is now mostly dry, with only occasional puddles of water.

The High Lakes Aquatic Alliance Foundation, a group based in Sisters, Oregon, suspects impacts of the ongoing drought are being exacerbated by ranching using Chewaucan River water between the Lake County ranching community of Paisley and Valley Falls, a small community near Abert's south end.

The Alliance, led by retired Oregon State University limnology professor Dr. Frank Conte of Sisters and retired U.S. Fish & Wildlife Service biologist Ron Larson of Klamath Falls, suspects Paisley area ranches are using the water at levels that threaten Abert's future. They also believe an earthen dam at the River's End Ranch, near Valley Falls, holds back water from flowing into Abert.

Figure 81. Photo by Lee Juillerat. A salt crystal packet removed from the former Lake Abert wet surface.

"Can't we get these people to realize water is important to everyone?" Conte wonders.

He's critical of the Oregon Water Resources Department for not monitoring the lower Chewaucan — the section from Paisley to Valley Falls. He's upset because the one-person brine shrimp harvesting business has shut down and, most of all, he fears a dry lake threatens migrating bird populations because "there's no alternative to this food source."

Conte claims lower Chewaucan irrigators have increased their water use over the years, a claim vigorously disputed by water users. And, even though those irrigators have water rights dating back to the 1800s, he says those rights may be superseded by international migratory bird treaties that he says guarantee the protection of migratory birds.

Asked if it's more important to maintain ranches that rely on the Chewaucan to irrigate fields for their survival, Conte replied, "It's more important to hold to the treaties."

Ranchers along the Chewaucan believe they're being unfairly targeted by the High Lakes group.

"What I point my finger at is the snowpack," says Mark Williams, manager of the ZX Ranch, one of the nation's largest ranches that has its headquarters south of Paisley. "You can have all the good moisture," he says, referring to precipitation that has fallen as rain instead of snow, "but if you don't have the snowpack …"

"We're all concerned about the outside chance of losing our water rights," says Sonny Simms, whose family ranch is south of Valley Falls. "The reality is we're in a severe drought. I don't believe any of us (irrigators) are using more water. I think we're actually more efficient. I think we are all trying to do everything we can to help fish and wildlife."

"We're using water that has been used for generations," echoes Joe Villagranna, who manages the J-Spear Ranch south of Paisley. "It's all drought driven. It's during the bad (low water) years you hear it's a problem."

"Our biggest concern is we want to make sure we can keep our water rights," echoes Martin Murphy, who oversees the multi-generation Murphy Ranch headquartered near Paisley. "It seems every time you turn around, people and new groups want to fight about something … The first thing they want to do is drag the ranchers down."

Murphy, Villagranna, Simms, Williams and others note the Chewaucan, as it now exists, is altered. Until the Lower Chewaucan was settled in the 1870s, the Upper flowed into a large marsh that extended south to the Valley Falls area toward Lake Abert. Diversion canals were created in 1916, effectively creating a river where none had existed. It's debated whether more, less or an equal amount of water from the "river" is used for irrigation than was lost to evaporation from the marsh.

Pete Schreder, an Oregon State University Lake County Extension agent and moderator of the Chewaucan Working Group, says historic records show water levels in Abert are cyclic — the lake was dry or nearly dry most years between 1926 and 1937 — and doubts water flowing into the lake would be greater if the river was still a marsh.

"The lake probably would still be dry," he believes. "Irrigation is not the issue."

PART 2 — Alliance blames irrigators for declining lake (Friday, September 11, 2015)

Yes, there is an ongoing drought and, yes, ranchers along the lower Chewaucan River between Paisley and Valley Falls have legally valid water rights to use river for irrigation.

But members of the High Lakes Aquatic Alliance Foundation think — and some insist — that ranchers are taking more water for irrigating fields and believe a taller dam built at the River's End Ranch, where the Chewaucan spills into Lake Abert, are factors in Abert drying up.

Because of concerns about Lake Abert, which dried up last summer and fared only slightly better this year, the Alliance, led by president Frank Conte, a retired Oregon State University limnology professor, sponsored a two-day symposium, "Lower Chewaucan Aquatic Ecosystem," in April at the Black Butte Ranch. He plans to publish a book on the symposium that will also feature magazine and newspaper articles about Lake Abert.

It's part of an effort by Conte and others to spotlight Lake Abert, a little-visited terminal, alkaline lake in northern Lake County that's regarded as a significant stopover for migrating waterfowl, with up to 80 species of water birds and peak numbers of an estimated 350,000 birds. Until it dried out, Abert produced brine shrimp, a favored food for migrating birds.

Although studies show periodic lake lowering and drying benefits brine shrimp production, and despite historic records showing the lake was mostly dry in the 1930s, Conte believes Lower Chewaucan water users are using more water, which has resulted in less reaching the Abert.

"I know they have," Conte says. "There is evidence to verify that."

He compares Lake Abert to Mono Lake, a terminal, alkaline California lake that nearly dried up in the 1980s and 1990s because water was sent to Los Angeles County to support and maintain its growth.

Sharing many of Conte's beliefs is Keith Kreuz, a former OSU student of Conte.

For Kreuz, the debate over Lake Abert is personal. For 35 years he was the person most intimately involved with the lake. He harvested brine shrimp, sometimes as much as 50,000 pounds a season, May through September through his business, Oregon Desert Brine Shrimp Company. He lost his business when the lake went dry in 2014. Although the lake has slightly more water this year, it was nowhere near enough to sustain brine shrimp populations that he estimates ranged from 10 million to in excess of 100 million pounds annually. That's a lot of shrimp — it takes about 8,000 brine shrimp to total a pound.

"It was basically a food factory," says Kreuz, 65, who lives in Portland, of Lake Abert.

During his early years, packaged shrimp was mainly sold as tropical fish food. As the business and demand grew, shrimp were sold to state and federal fish hatcheries and to companies that produce food for family fishbowls. In the past 10 years, Kreuz says there was a virtually unlimited demand from overseas markets who used the shrimp as feed on fish farms in Indonesia, South America and elsewhere. Unlike the brine shrimp from China, the Abert Lake variety are known for being uncontaminated, disease-free and free of pathogens.

"We were just on the verge of making something happen," he says of meeting the increasing demands. Kreuz said he hesitated from expanding his business because he saw water levels decrease over the past decade. "I've seen that water drop at an unprecedented rate."

While he acknowledges the multi-year drought, he believes the decline is the result of increasing upstream demands from ranchers and others who use water from the Chewaucan River, the main source of Abert Lake.

"It is my belief that the lake is currently dry because of drought over the past four years and with increased water usage upriver," he says, noting three independent studies "all point to the same conclusion — that the drying of Abert Lake cannot alone be attributed to drought conditions."

Kreuz says people who believe the lake's declining water level is drought driven "have produced no empirical data or study to support their position. It must be remembered that Abert was reaching near lethal and low levels by 2010, which is before the drought we have experienced these last four years. It only seems reasonable that the increased water use over the past 20 years plus has had a cumulative effect with each additional withdrawal. When one considers the water lost by the Rivers End Reservoir and OWEB's (Oregon Watershed Enhancement Board) flooding of 3,000 acres in the upper Chewaucan marsh, along with other withdrawals, one can see why flow into Abert has declined."

He believes the relationship between lowering the water table and river flow needs to be considered, asserting, "All one needs to do is drive from Paisley to Valley Falls and then out to Lake Abert and you can see more and more land being tilled and put into production." He says sagebrush fields have been replaced with alfalfa, a crop that uses two to three times more water than hay.

Although Lake Abert went dry during the 1930s, Kreuz says scientific literature indicates that dry period was an anomaly, pointing to studies that show a similar severe drought has not been experienced in the region for 700 years. He believes the dry conditions may continue.

"I am quite certain that if the present water usage upriver continues, the frequency of a dry lake will be much more common and will likely be dry or in a near dry condition, even during periods of slightly lower than average precipitation," he says. "This story has played itself out in almost every other lake in eastern Oregon, whether it be Silver Lake, Summer Lake, Goose Lake. The frequency of these lakes being dry has increased greatly when compared to times before there was excessive river water withdrawal."

Kreuz says he's passionate about Lake Abert "not only because I have lost everything I have put into my company for 35 years but, more importantly, the loss of a rare and unique ecosystem that supports 3.5 million migratory birds — a public treasure that everyone has ownership in, and not just those with agricultural interests. I do understand and sympathize with the plight of the local rancher/hay producer, but I only wish similar concern and understanding be brought to the plight of this unique and valuable lake."

He also believes another factor figures in — "The longer Lake Abert is low or empty, the more people will say this is the new normal, so people will forget about it."

* * *

Lake Abert facts — By Ron Larson and Frank Conte

• Lake Abert is Oregon's sixth largest lake and its only highly saline lake. Its saline waters and extensive mud flats provide habitat for migrating shorebirds in the spring and fall.

• Brine shrimp and shore flies are the only common aquatic species in the lake and can occur in high numbers. Brine shrimp are quarter-inch-long primitive crustaceans that are not actually shrimp and live only in high-salinity, fishless lakes. They are easily caught by water birds near shore and in open water. Adult shore flies can be so numerous the lake shore sometimes looks covered by tar. Their larvae that live in the mud are easily caught by birds.

• Up to 80 species of water birds have been reported. Total bird numbers peak in late July and early August, then decline through fall. In July 2013, water birds reached an estimated 350,000.

• Phalaropes are Lake Abert's most numerous water birds, with numbers reaching more than 300,000, about a fifth of the world population. The sandpiper-like birds are about the size of a red-winged blackbird. They are unusual because the female is the most colorful and compete for males, and the male does all the nesting. Phalaropes nest in wetlands and migrate to Abert to molt, stage and feed — they double their body weight — before migrating to South America, a 50-hour, non-stop, 5,000 mile flight. They over-winter in salt lakes in the high Andes of Peru, Chile, Bolivia and northwest Argentina.

• State-listed threatened Snowy Plovers nest at the lake.

• The only thing that can break the cycle of increasing salt load is almost complete drying of the lake and the removal of sale precipitates by blowing wind. The periodic drying of the lake maintains the lake's long-term productivity because of lower salt levels.

• Lake Abert was thought to be dry in 1924, 1926, 1930, 1931, 1933, 1934 and 1937.

• The Chewaucan Marsh, between Paisley and Valley Falls, was drained between 1884 and 1915, and most of the land was converted to agricultural use and some of the waters were used to irrigate farm and ranch lands. By 1965, water lost from Lake Abert due to irrigation was about equal to water lost from evaporation from the Chewaucan Marsh.

PART 3 — Why is Lake Abert Drying? Many say drought (Saturday, September 12, 2015)

For many people, the reason Lake Abert is drying up is no mystery.

"Most of Lake Abert's water comes from spring snow melt," says Craig Foster, a biologist with the Oregon Department of Fish and Wildlife's Lakeview office. "When you don't get a good snowpack in the mountains, the water drops. Anytime you get two drought years, the level drops."

Foster notes long-term weather records indicate the average precipitation near the lake is 10 to 11 inches a year, and that evaporation is 45 inches annually.

"It is what it is," he believes of Abert's historically cyclic lake elevation variations. "It's been dry before. It's a drought-driven system."

Foster recognizes Abert's importance to hundreds of thousands of migrating birds, "But we're in a drought. That's not the answer they want to hear," he says, referring to the High Lakes Aquatic Alliance, WaterWatch of Oregon and other groups that want water from the lower Chewaucan River between Paisley and Valley Falls sent to Lake Abert during low precipitation years. Water used for irrigating ranches, which provides habitat for wildlife and a variety of birds, would be severely reduced.

He's not the only one who questions the Alliance's proposal. Pete Schreder, an Oregon State University Lake County Extension agent and moderator of the Chewaucan Working Group, also wonders about impacts on waterfowl and wildlife along the lower Chewaucan that benefit from irrigated fields.

"It's a larger ecosystem than Lake Abert," Schreder says, rhetorically asking, "If we dry up the river and put water in the lake, what is the impact?"

Alliance members and officials with WaterWatch also question why Chewaucan water is impounded behind a dam at the Rivers End Ranch, where the river empties into the lake. In a recent newsletter article, "WaterWatch," Jim McCarthy, southern Oregon program manager, calls the reservoir and dam "a key contributor to the lake's rising salinity" and claims no protective measures, such as a freshwater bypass, were required when the Oregon Water Resources Department permitted a taller dam in 1991.

"Rivers End Ranch," McCarthy writes, "originally approved as a cooperative wildlife and wetland restoration project and constructed with significant public funding, now operates free of requirements to protect the lake and without the intended wildlife agency oversight."

Foster, however, disputes those claims. He also says studies indicate that draining the Rivers End reservoir, which provides habitat for fish and waterfowl, would raise Abert's water level two-tenths of an inch and evaporate in three hot summer days. He also says sending water down the Chewaucan River, a man-made river that replaced shallow wetlands, would provide water for only a few days before being lost to evaporation.

"The Chewaucan system from Paisley downstream is extremely important," he emphasizes, noting the river and periodic wetlands created by irrigation provides habitat for snow and white-fronted geese, sandhill cranes, egrets and other birds.

"It's a drought. We have a small percentage of what we normally have," says Brian Mayer, Lake County's water master, who notes the county issued an emergency drought declaration in March. "They could take a lot of water but they don't," he said of lower Chewaucan irrigators.

Likewise, Kyle Gorman, Oregon Water Resources regional manager from Bend, believes irrigators are "unduly blamed," noting, "It's fairly obvious it's been a dry cycle ... In my professional experience I do not believe there has been more water use than is historical." He termed the impact of water storage at the River's End Ranch as "very minimal."

Lake County commissioners Dan Shoun and Ken Kestner adamantly dismiss claims irrigators are to blame for Abert being nearly dry.

"Ridiculous. All of it is speculation by outside interests," says Shoun, who lives in Paisley, of claims that irrigators are to blame. "The sooner we get that dispelled and protect our ranching community from outside interests, the better. People are not going to turn off the water and let it flow to the lake."

"We're in a severe drought," Kestner says, noting Abert has been dry other years, something that's beneficial because it helps disperse salt created by the lake's alkaline waters. "Nature is not consistent year after year."

He notes the Lower Chewaucan formerly was a marsh that was dredged to create a "river" and believes "the amount of water used for irrigation is no greater than what would have evaporated."

Kestner believes ranchers "are very conscious about the environment and look for efficiencies." He believes those efficiencies are practiced by the several multi-generational family ranches and the ZX Ranch, one of the nation's largest ranches that is owned by the J.R. Simplot Company, one of the world's largest privately owned companies.

"The bigger operations have more ability to be environmentally conscientious," Kestner says. "The larger ranches have the flexibility and adaptability to safeguard the environment."

Ron Larson, a retired U.S. Fish and Wildlife Service biologist and member of the High Lakes group, believes greater cooperation is needed between irrigators, state officials and people concerned about the long-term impacts on Lake Abert. He says the lack of data makes it impossible to know how to compare the amount of water taken from the Lower Chewaucan in recent years compares with historic losses to evaporation.

"It's pretty clear the hydrology has been altered," he says. "There are so many straws in the river … It's a cumulative effect. How do you balance the upstream versus the downstream needs?"

For several years, Larson has been collecting data on Lake Abert levels. He visits the lake and hikes to an overlook to photograph and document lake changes. While studies show that low lake levels can benefit the lake's ecosystem, he warns, "If the lake stays too dry for too long, it upsets the whole cycle. If it's irregular for too long, it's hard for any organism to adapt." While he notes brine shrimp can stay in the lake sediment during dry times, "They don't last forever."

Larson wants Oregon Water Resources to become more involved by monitoring the lake and installing flow meters at all diversions. He also believes the state should consider water as a resource like petroleum, and be subject to user fees and monitoring. "I think they realize they have it pretty good," he says of irrigators. "What other resource does anyone get free for life?"

Larson believes some problems, and distrust,

Figure 82. Photo by Larry Turner. Joe Villagrana who manages the J-Spear Ranch south of Paisley believes water concerns in the Lower Chewaucan River and Lake Abert are drought-driven.

can be resolved. "What I would appreciate from ranchers is a process to get the data ... to make it a very transparent process."

Villagranna, however, says those steps are being taken and believes critics need to visit ranches to understand

Figure 83. Photo by Larry Turner. Mark Williams, manager of the ZX Ranch near Paisley, along with other area ranchers, believes they're being unfairly targeted by the High Lakes Aquatic Alliance Foundation.

their different, often complex water management operations. "We'd all be in favor of visiting with them (environmental groups), but we've done nothing wrong." He says computerized measuring devices are being installed so "we'll know what each ditch is pulling out. All that water that's going out of the river will be accounted for."

He emphasizes the information isn't designed to placate critics but is intended to help irrigators better manage their water allocations. Villagranna used this year's drought forecasts to modify the J-Spear's operations. "We've had to change how we irrigate," he explains, noting by late August irrigation water was being used on only about 100 of the ranch's 2,000 acres along the Chewaucan. Earlier in the year, again using drought forecasts, he decided to wean and sell calves earlier than usual and not carry over as many cows through the winter.

Other ranchers share Villagrana's concerns.

"We're irrigating pretty much the same ground Mother Nature irrigated 200 years ago," says Mike O'Leary of the O'Leary Ranch. He believes the media generally favors environmental groups and issues, so "We're trying to tell our side of the story. When we turn our water on, the birds, you just can't count them. Everything we've done on the Chewaucan and the diversions are for wildlife and fish. I don't like to see Abert dry either, but it's natural."

"There's not a rancher who wants to see the lake go dry," agrees Mark Williams, the ZX's ranch manager. "Mother Nature is our director, and

from there we all use water in the most beneficial way we can. I hope calm heads and calm thinking prevails."

"The bottom line comes down to we need winters," echoes Sonny Simms, whose family ranch is south of Valley Falls. "The solution for all of us would be for Mother Nature to cooperate and give us some good winters."

Figure 84. Photo by Lee Juillerat. Sonny Simms, whose family's ranch is south of Valley Falls, believes irrigators are being more efficient with water use in the face of severe drought conditions.

Until those winters return, the debate

Figure 85. XL River Ranch (see OPB video Episode 2610 for Lower Chewaucan River Shoreline and Marsh details).

will continue. As Ron Larson says, "We all have different value systems." Asked his thoughts on what's more important, channeling water to Lake Abert to improve the odds that lake levels will rise so that brine shrimp might be available for migrating waterfowl, or leaving the system intact for ranching families and others who depend on Chewaucan River water for irrigation and other uses, Larson hesitates before replying, "That's something people need to decide for themselves."

Figure 86. View from the southeastern pasture on Rivers End Ranch adjacent to Lake Abert.

Health/Medical Industry—Migratory Avian Influenza Virus
by F. P. Conte

Avian Influenza Virus (HPAF1)

The emergence of a highly contagious avian virus carried by the migratory bird population has caused a worldwide health problem. In the United States, the states of Iowa and Minnesota have suffered major financial disasters in the poultry industry. The chicken, duck, and goose industry in Iowa had to sacrifice 29 million birds, while in Minnesota the sacrifice was 8 million birds. It has been estimated that 200 farms in fifteen different states have lost a total of more than 45 million birds.

Has the State of Oregon suffered any losses as a result of this virulent and pathogenic virus? Yes, in Oregon it was first found in mid-December in a flock of backyard birds in Winston, in Douglas County. Another crop of domestic birds has been found in Deschutes County in a flock of ninety chickens, ducks, and turkeys on a farm near the town of Tumalo, Oregon. This farm had access to ponds that were frequented by migratory wild waterfowl. As of February 2015, the current assessment of the distribution of avian flu outbreak, according to a report by the Oregon, Washington, and Idaho Departments of Agriculture, is that eight backyard poultry flocks and three captive wild birds have contracted the virus, including a domestic turkey farm in California. So far, these states' agricultural and health officials have not declared that an avian flu epidemic has occurred. However, the Oregon Department of Fish and Wildlife has been asking people to report wild bird deaths by calling their advisory board in Salem, Oregon, and have been advising people to avoid contact with sick or dead wild or domestic birds.

Health Officials and Wildlife Managers

Efforts are continuing at both state and federal levels to test wild birds for all strains of the avian influenza virus found in the western states that have habitats for migratory waterfowl. This is an example of good environmental and financial management of our natural watershed resources.

A recent publication by Norrie Russell, of the Roslin Institute, University of Edinburgh, discussed avian research dealing with domestic chicks that were genetically modified in a bid to block bird flu. In their early experiments they discovered a compound they considered promising in fighting off the disease that has devastated poultry in the U.S. In these experiments, researchers injected chicks with fluorescent protein to distinguish them from normal birds (see Figure 87).

Health regulators around the world (including Oregon) have yet to approve any animals bred as genetically modified organisms (GMO) for use in the food industry, because of long-standing safety and environmental concerns.

Figure 87. Baby chicks modified with a "decoy" gene. By tracking their progress, the experts can monitor how susceptible they are to bird flu. Ultraviolet light in the beaks and leg segments show proteins blocking the flu.

Water Balance Hydrology of Lower Chewaucan River (Water In = Water Out)

Past Modeling

The model that gave this data was presented by George Keister (1992, 2014). Figure 88 shows, via a comparison of yearly lake levels from 1928 through 1986, that historically there has been sufficient water inflow to keep Lake Abert at a functional level. The curves are derived from water surface/volume computed by this model as water outflow from the Chewaucan River into Lake Abert, and the measured water inflow as measured at the city of Paisley, Oregon. You will note that both the computed and measured lake levels form a nearly identical curve. The interpretation given this data is that measured freshwater inflows at Paisley, Oregon, could predict the volume of water outflow that should be reaching Lake Abert. This freshwater inflow would offset any impacts produced by climatic drought or agricultural diversion permits issued by the Oregon Department of Water Resources during this period. At the time of the report, the hydrology predictions for Lake Abert for both dry and wet periods appeared to be nearly identical, and therefore accurate within a degree of five to eight percent. This data was collected at a time when the Chewaucan marsh had been in existence for 100 years, since well before the current agricultural ranches/farms started operating with their high-hydraulic sprinkling systems. Despite the variations in lake levels during several dry periods (1927, 1930/31, 1933/34, 1937), the model system depicts the dry periods as 1932, 1933, and 1934, which compares with the earlier hydrological predictions of closed lakes in the western Great Basin by Harbeck (1955; Hasting, S. T. 1965) and closed basins in the Lower Chewaucan River Basin by Philips and Van Denburgh (1971).

Current Modeling Technology—DOGAMI–LIDAR
Measurements of Surface Area/Water Volumes

The 2010 Oregon Department of Geology and Mineral Industries (DOGAMI) Light Detection and Ranging (Lidar): Klamath Study Area and Lake County used radar technology and aerial flights to establish Lidar with airborne systems to generate precise three-dimensional information about the shape of the Earth and its surface characteristics.

When an airborne laser is pointed at a targeted area on the ground, the beam of light is reflected by the surface it encounters. A sensor records this reflected light to measure a range. When laser ranges are combined with position and orientation data from integrated geophysical systems, and with scan angles and calibration data, the combination of data points is rich in other detail-rich groups of elevation points. This becomes what is termed "a point cloud." These point clouds are used to generate other geospatial products, such as canopy models, building models, and contours of Earth or volumes of lakes. The use of Lidar on Abert Lake to determine its volume in both wet-drought conditions and climatic-dry conditions may supply the water-outflow measurements needed by OWRD hydrologists to determine whether the freshwater flow of the Lower Chewaucan River system is sufficient, at its point of entrance into the lake, to maintain the health and nutrition of the lake's biotic inhabitants. It is hoped by members of the non-profit High Lake Aquatic Alliance Foundation that the organization's goals for Lake Abert's aquatic ecosystem can eventually be met.

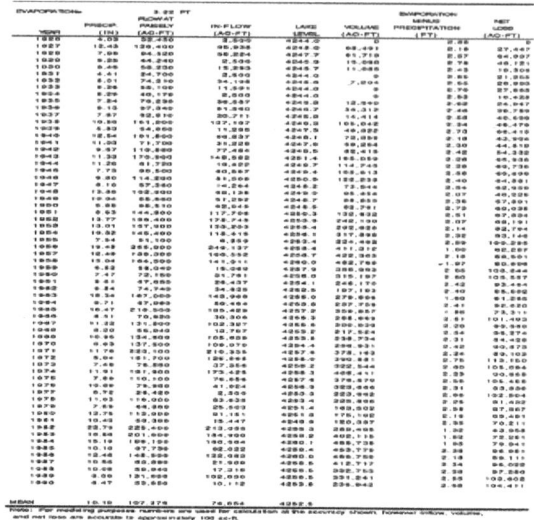

Figure 88. (From Keister, 1992)

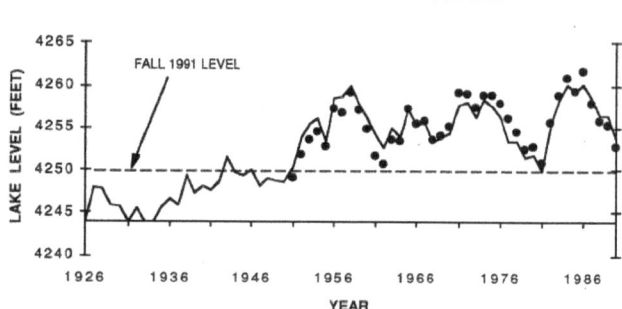

Figure 89. Lake level. (Keister, 1992)

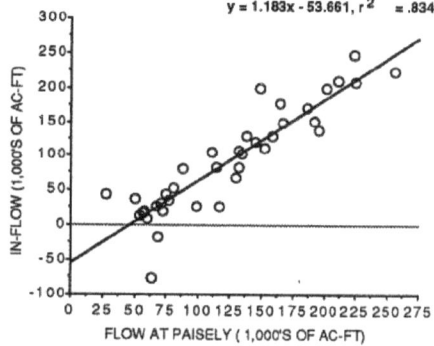

Figure 90. Paisley flow. (Keister, 1992)

Summary of Hydrologic Studies

Report	Study Period (Water Years)	Average Chewaucan Flows @ Paisley (TAF)	Average Inflows to Lake Abert from the river (TAF)	Percent of Chewaucan River flows @ Paisley	Percent of Total Inflows from Chewaucan River
Phillips and VanDenburgh 1991	1951-1962	98	48	51	
VanDenburgh 1975	1916-1965	96	62	65	90
Keister 1992	1926-1990	107	75	70	
Larson unpub	1974-2014	114	66	45	66
	2000-2014	110	43	35	62

Table 4. Summary of Past and Present Hydrological Data.

APPENDIX A
Photographs of Conference Participants and Audience

by Conrad Weiler, HLAAF Photographer

The following pictures are of the presenters and audience members who participated in the conference at Black Butte Ranch in the Symposium, April 25, 2015.

Speakers at the start of the conference

APPENDIX B
Conference Posters, with Photographs of Speakers and Abstracts

by Dee Ann Glazer, Glazer Graphics and Illustrations, Bend, Oregon

History of Brine Shrimp Fishery in Lake Abert and the role in its Ecosystem

Frank P. Conte, PhD
Professor of Zoology (Emeritus)
Oregon State University

Keith Kruez, MS
Fisheries Graduate/Owner
Oregon State Univ./ ODBS

ABSTRACT: In 1970, Prof. Conte made the discovery that in the brine shrimp (Artemia), during its embryonic development from a dried cyst that, when it became hydrated from fresh rain water, and slid into its home in saline lake water, the larval nauplius formed a new salt gland. This gland enable the shrimp to gulp algae/salt water and digest the algae for energy to grow into adults and at the same time to save water for osmotic and ionic balance of its internal fluids. From 1970 to 1975, the shrimp model became the ideal model for scientists studying how the ion-pumps were formed from molecular genes. Many graduate students from all over the USA came to OSU to learn how the shrimp could live in saline water. One was David Herbst, who worked with FPC and son Paul at Mono Lake studying the brine shrimp in 1976-77 and came to OSU in 1977-78 to study brine shrimp and brine flies under FPC. At this time FPC wanted to get a different brine shrimp for bimolecular genetic studies and learned from Dave that Lake Abert contain a large number of adult brine shrimp and many cysts lined the shores. FPC confirmed this information in 1978 and in 1979 wrote an OSU Sea Grant Proposal. Discussions with other state agency officials was both numerous and hot and heavy with critics. One was a 1979 master's degree graduate named Mr. Keith Kruez. Keith and his colleague Greg Brown had earlier in 1979 began monitoring the location and density of brine shrimp population in the lake and found that it indeed could support a brine shrimp fishery. They began building a floating platform that could be attached to large zooplankton nets and would catch large numbers of adult shrimp. In addition, they developed cleansing stations along the shoreline to clean the shrimp of excess salt and algal debris. In the small community of Valley Falls, they built the machiney to package the shrimp into frozen packets of different sizes and stored in frozen storage facility. Thus began, the Oregon Brine Shrimp Company, whose headquarters is in Portland, Oregon. Keith Kruze will tell you about building the only known commercial enterprise utilizing the resources of Lake Abert without injury to its ecosystem for the last 35 years.

The Oregon Public Trust Doctrine and Lake Abert

Michael C. Blumm
Jeffrey Bain Faculty Scholar & Professor of Law
Lewis and Clark Law School

ABSTRACT: Professor Blumm will discuss the state of Oregon's public trust doctrine, its origins in the Oregon Statehood Act, and recent cases concerning the atmosphere and Oswego Lake as trust resources. Although there is a fair amount of recent litigation, Oregon's doctrine has never been applied to water diversions. California's doctrine has, however—most notably in the California Supreme Court's famous Mono Lake decision in 1983. Application of the trust doctrine to Mono Lake, a terminal saline lake not unlike Abert Lake, has led to considerable lake and stream restoration in the Mono Lake Basin. Although the two lakes are only about 300 miles apart, Abert Lake has the misfortune of being located in Oregon, just 50 miles or so from the California state line. Professor Blumm will discuss the prospect of bringing California's interpretation of the public trust doctrine to Oregon, a development which could potentially save Abert Lake.

The Changing Weather: An Overview of Potential Climate Change Impacts on Central Oregon

Stephen W. Bieda III, PhD
Meteorologist, Acting Science & Operations Officer
NOAA National Weather Service, Pendleton, OR

ABSTRACT: The Fifth Assessment Report (AR5) by the Intergovernmental Panel on Climate Change (IPCC) reported that there is strong scientific evidence suggesting a human influence on the global climate. One of the most noteworthy changes being observed in the climate record includes a tendency towards warmer overall temperatures, and increased occurrences of extreme weather or climate events. Weather and climate data for central Oregon is explored to assess past and present trends of temperature and precipitation. A closer look at the conditions over the winter of 2014 and 2015 will also be explored due to the unseasonably warm temperatures that were accompanied by average precipitation and little snowfall. A brief overview on climate change be provided concerning the latest climate model projections and potential ramifications for Central Oregon.

Lake Abert – "An Uncertain Future for Oregon's Internationally-Important Migratory Bird Habitat and Only Saline Ecosystem"

Ron Larson, PhD

ABSTRACT: In 2014, when all of the lakes in south-central Oregon were dry, the event was barely noticed by the news outlets, nor did State officials alert the public to the event. However a small group of concerned citizens collected data to document the event. The focus was on Lake Abert, Oregon's 5th largest lake, and Oregon's only saline ecosystem, being many times saltier than the ocean. The lake was once part of a much larger lake that shrank about 10,000 years ago and was the home of Native Americans who hunted and fished around the lake and at nearby marshes. Although Lake Abert is recognized as being internationally important for migratory waterbirds, attracting hundreds of thousands every year; it has no protection and consequently its future is uncertain because of climate change and unsustainable water use by agriculture.

Brine flies on the rise: the natural history, ecology, and physiology of the brine fly *Ephydra hians*

Gregor Yanega, PhD
Academic Coordinator of the Salton Sea Initiative
University of California, Irvine

ABSTRACT: In the evolutionary history of arthropods, adaptations to living in salt water are rare. Brine flies (Diptera:Ephydridae) are widespread and abundant in saline lakes. In this review, I discuss the physiological and morphological aspects of brine flies that allow them to succeed in saline lakes and comment on their ecological importance as consumers of algae and as prey to migratory birds. Morphological and physiological modifications of the hindgut have made osmoregulation in saline environments possible, and the development of lime gland lined with calcium-carbonate allow larval flies to remain submerged despite being otherwise positively buoyant. I conclude by considering the role of brine flies in the food chain of saline ecosystems with and without fish.

Waterbird Responses to Semi-Arid Land Wetlands Altered as a Result of Changing Climate

Susan M. Haig, PhD

Senior Scientist, USGS Forest and Rangeland Ecosystem Science Center
Professor of Wildlife Ecology, Oregon State University

ABSTRACT: Anthropogenic climate change is altering aquatic ecosystems worldwide, particularly small and shallow wetlands. As a result of these abiotic changes, the terrestrial and aquatic species that depend on such wetlands are also likely to experience significant shifts in range, phenology, and population structure, particularly in arid and semi-arid regions that are already limited in water quantity and quality. We are developing a framework to determine and manage landscape-level impacts of climate change on wetlands and wetland-dependent species in semi-arid areas of North America's Great Basin. We begin describing the scope of abiotic impacts from climate change using remote sensing, historical records, and ground-level monitoring to create models of the relationships between water volume, water quality, weather, and climate. We can use projections of future climate conditions to model how wetland habitat quality and species connectivity will change in the coming decades by combining estimates of landscape connectivity for these species with a model of the climate drivers of wetland patch quality. This approach will serve as a general model for understanding population- and community-level climate impacts and provide a sound basis for conservation planning and adaptive management by wetland resource managers around the world. The model will be made user-friendly for specific wetland management as well as provide regional perspectives.

Importance of Highly Saline Lakes to Eared Grebe Life History

Annette Henry, Survey Coordinator
Marine Mammal and Turtle Division
NOAA Fisheries

ABSTRACT: The Eared Grebe (*Podiceps nigricollis*) is the most abundant grebe in the North America. It is a visual predator and pursues its prey underwater, eating mostly small crustaceans, aquatic insects, and larvae. It prefers to swimming to flying. Indeed, the Eared Grebe is essentially flightless for most of the year and is one of the most inefficient fliers among birds; flight is almost exclusively used for long-distance migration. Eared Grebes make a postbreeding/molt migration from breeding areas in the interior of western North America to hypersaline lakes in the Great Basin and the majority of birds migrate between early August and mid-October. It is at these lakes the birds molt and stage for several months to prepare for their subsequent migration to wintering areas south. After arrival to the saline lakes, the birds undergo the most dramatic change in body composition known in birds. Breast and heart muscles atrophy, and birds become flightless. At the same time, the birds increase the size of their digestive organs (stomach, liver, intestines) and put on large fat stores. The birds more than double their arrival weights. And then shortly before departure to their wintering grounds, these changes are reversed and the birds catabolize fat stores, increase heart and breast muscle size, and reduce their digestive organs for the southbound migration. The Eared Grebe is one of a few species that is able to exploit the superabundant prey that thrive in the highly saline lakes, brine shrimp and alkali flies. Hypersaline lakes are few and far between in North America and are critical habitat for the Eared Grebe. There rarity makes each one an exceedingly valuable resource.

The Importance of Western Great Basin Wetlands to Waterbird Connectivity Throughout the Annual Cycle

Lewis W. Oring, PhD

Professor of Natural Resources (Emeritus), University of Nevada

ABSTRACT: Wetland conservation plans generally focus on single sites and rarely consider the diversity of water types and other ecological needs by the myriad of waterbirds using them throughout the annual cycle. In dynamic areas such as the Great Basin, location of the most important wetlands in North America, conservation efforts neglecting to consider these factors will fail to provide adequate habitat for an enormously rich avifauna. For example, at different times of the year, hypersaline Lake Abert hosts most of the world's Wilson's Phalaropes, Eared Grebes, American Avocets, etc. Our research focused on learning how shorebirds used various types of wetlands throughout the year across the basin. We found radio-tagged American Avocets would fly 100-300 km after they had finished breeding to stage at hypersaline Lake Abert. Conversely, Killdeer tended to remain within 1 km of their freshwater breeding territory all year. Intermediate in space-use, radio-tagged Willets nested near freshwater breeding sites but foraged in saline areas and left the basin for the coast as soon as their chicks fledged. Thus, understanding the behavior of just three species illustrates the diverse needs of birds and importance of the mosaic of Great Basin wetlands.

References

Books

Conte, Frank P., Craig Conte, and Paul Conte. 2012. *The Biological Secrets of Salt: Its diversity in Organisms and Impact on Humans.* http://www.amazon.com/Biological-Secrets-Salt-Diversity-Organisms/dp/1480179604/

Conte, Frank P., Craig Conte, and Paul Conte. 2014. *The New Biological Secrets of Salt: Its Diversity in Organisms and Impact on Humans.* http://www.amazon.com/New-Biological-Secrets-Salt-Diversity/dp/1503055922/

Danielli, J. E., ed. 1984. *International Review of Cytology, Vol. 91.* Chapter 2: F. P. Conte, "Structure and Function of the Crustacean Larval Salt Gland," pp. 45–104. Orlando, Florida: Academic Press, Inc.

Jeon, K. W. 2012. *International Review of Cell and Molecular Biology.* Ed., Chapter 1, F. P. Conte, "Origin and Differentiation of Ionocytes in Gill Epithelium of Teleost Fish," pp. 1–27. London: Elsevier.

———. 2008. *International Review of Cell and Molecular Biology.* Ed., Chapter 2, vol. 268, F. P. Conte, "Molecular Domains in Epithelial Salt Cell (NaCl) of Crustacean Salt Gland (Artemia)," pp. 39–57. London: Elsevier.

Jones, J., et al. 2015. *California's Most Significant Droughts: Comparing Historical and Recent Conditions.* State of California: California Department of Water Resources, pp. 1–128.

Keister, G. 2014. *A Look Back at Lake Abert and The River's End Project.* Special Report Addendum to 1992 OFW (unpublished).

———. 1992. *Ecology of Lake Abert: Analysis of Further Development.* ODFW Special Report (unpublished).

Lahlou, B., ed. 1980. *Epithelial Transport in the Lower Vertebrates, Part III.* Chapter 4: F. P. Conte, "The Ionocyte Chalone: A chemical regulator of chloride cell proliferation and transformation," pp. 287–296. London: Cambridge University Press.

Larson, G., and R. Larson. 2011. Featured Lake, Lake Abert, NALMS Winter, pp. 47–58.

Patten, D. T. 1987. *The Mono Basin Ecosystem, Effects of Changing Lake Level, Mono Basin Ecosystem Study Committee.* Frank P. Conte. Washington, D.C.: National Academy Press, pp. 1–213.

Waynes Word—California Pink Salt Lakes. Photos by W. P. Armstrong.

W. P. Armstrong. 1981. "The Pink Playa of Owens Valley." *Fremontia* 9:3–10.

Publications and Scientific Journals

Biewener, A. 2011. Vol. 366:1570, Published 18 April 2011. DOI: 10.1098/rstb.2010.0353)

Blinn, D. W. 1971. Autoecology of a filamentous alga, *Ctenoclatus,* in saline environments. *Can. J. Biol.* 49:735–743.

Boula, K. M., and A. L. Jarvis. 1984. Foraging ecology of fall-migrating waterbirds, Lake Abert, Oregon. Corvallis: Oregon State University. 29 pp.

Bowen, William S. 2003. *Oregon Atlas of Panoramic Aerial Images.* Online at geodata.csun.edu/ or-panorama.atlas/index.html

Cannon, W. J. 1977. Cultural resources survey of west shore of Lake Abert. U.S. Dept. of the Interior, Bureau of Land Management, Lakeview District Office. Report No. 010-12-78/12-28-77

Clegg, J. S. 2011. Stress-related protein compared in diapause and in anoxic encysted embryos in diapuae and in anoxic encysted embryos of *Artemia fransciscana. J. Insect Physiol.* 57:660–664.

Cole, D. L., and R. M. Pettigrew. 1976. Archaeological survey of the proposed improvement of the Pikes Ranch—Valley Falls Section, North Unit of the Lakeview-Burns Highway, Lake County, Oregon. Eugene, Oregon: Museum of Natural History. 25 pp.

Conte, F. P., and D. Lin. 1967. Kinetics of cellular morphogenesis in gill epithelium during and under water adaptation of *Ocorhynchus. Comp. Biochem. Physiol.* 23:945–957.

Conte, F. P., and P. A. Conte. 1988. Abundance and spatial distribution of *Artemia salina* in Lake Abert, Oregon. *Hydrobiologia* 158:167–172.

Gill, T. E., and D. A. Gillette. 1996. Owens Lake: A natural laboratory for acidification, playa dessication, and desert dust. *Geomorphology* 17:229–248.

Grayson, D. 1993. The Desert's Past: A natural prehistory of the Great Basin. Washington D.C.: Smithsonian Institution Press. 384 pp.

Harbeck, G. E, Jr. 1955. The effects of salinity on evaporation: U.S. Geological Survey of selected Western lakes and reservoirs for water-loss studies. U.S. Geol. Survey Ctr. 103. 31 pp.

Harding, S. T. 1965. Recent variations in the water supply of the Great Western Basin. University of California, Berkeley: Water Records Center Archives.

Herbst, D. B. 1990. Distribution and abundance of the alkali fly (*Ephydra hians*) Say at Mono Lake, California (USA) in relation to physical habitat. *Hydrobiologia* 197:193–205.

Herbst, D. B., and T. J. Bradley. 1989. Salinity and nutrient limitations on growth of benthic algae from two alkaline salt lakes of the Western Great Basin (USA). *J. Phycol.* 25:673–678.

Herbst, D. B., F. P. Conte, and V. J. Brookes. 1988. Osmoregulation in an alkaline salt lake insect, *Ephydra* (Hydropyrus) *Hians* Say (Diptera: Ephydridae) in relation to water chemistry. *J. Insect. Phycol.* 34:903–909.

Herbst, D. B. 1988. Comparative population ecology of *Ephydra hians* Say (Diptera: Ephydridae) at Mono Lake (California) and Abert Lake (Oregon). *Hydrobiologia* 158:145–166.

―――. 1986. Comparative Studies of the Population Ecology and Life History Patterns of an Alkaline Salt Lake Insect: *Ephydra Hians*. Doctor of Philosophy dissertation. Corvallis: Oregon State University. pp. 1–206.

―――. 1981. Ecological physiology of the larval brine fly (*Ephydra*). Master's thesis. Corvallis: Oregon State University. pp. 1–61.

Hildebrandt, J., R. Gerstburger, and M. Schwartz. 1998. In vivo and in vitro induction of C-FOS in avian exocrine gland. *American Journal of Physiol.* 275:C951–C957.

Hoffman, D. J. 2002. Role of selenium toxicity and oxidative stress in aquatic birds. *Aquatic Toxicology* 57:11–26.

Hossler, F., M. Sarras, and F. Allen. 1978. Ultrastructural, Cyto- and Biochemical observations during turnover of plasma membrane in duck salt gland. *Cell. Tiss. Res.* 188:299–315.

Hunt, P., H. E. Kilham, J. M. Klieforth, J. M. Melack, and S. A. Temple. 1987. The Mono Basin ecosystem. Washington, D.C.: National Academy Press. 272 pp.

Hwang, P. P., T. H. Lew, and L. Y. Lin 2011. Ion regulation in fish gills: recent progress in the cellular and molecular mechanisms. *Am. J. Physiol. Regul. Integ. Comp. Physiol.* 301:R28–R47.

Hwang, P. P., and M. Y. Chou. 2013. Zebrafish as an animal model to study ion homeostasis. *Pflugers Arch.* 465:1233–1247.

Jehl, J. A., Jr. 1988. Biology of the eared grebe and Wilson's phalarope in the nonbreeding season: A study of adaptation to saline lakes. Pp. 46–71 in F. A. Pitelka, ed. *Studies in Avian Biology No. 12.* University of California, Los Angeles: Cooper Ornithological Society.

Knight, C. K., and M. Peaker. 1978. Adaptive hyperplasia and compensatory growth in the salt glands of ducks and geese. *J. Physiol.* 294:145–151.

Kristensen, K., M. Stern, and J. Morawski. 1991. Birds of North Lake Abert, Lake Co., Oregon. Oregon Birds 17 Central.

Langbein, W. B. 1961. Salinity and Hydrology of Closed Lakes, Paper 412. Washington, D.C.: Geological Survey Professional Society Papers.

Mahoney, S. A., and J. A. Jehl, Jr. 1987. Avoidance of salt loading by a diving bird at a hypersaline and alkaline lake, eared grebe. *Condor* 87:389–397.

Morawski, J., and M. Stern. 1991. Lake Abert Water bird counts. Unpublished report. The Nature Conservancy. 9 pp.

Patten, D. T., F. P. Conte, W. E. Cooper, J. Dracup, S. Dreiss, K. Harper, G. L.

Petterson, G. L., R. D. Ewing, S. R. Hootman, and F. P. Conte. 1978. Large-scale partial purification and molecular and kinetic properties of Na+K-activated adenosine triphosphatase in Aremia. *J. Biol. Chem.* 253:4762–4770.

Pettigrew, R. M., and Albert C. Oetting. 1985. An Archaeological Survey in the Lake Abert-Chewaucan Basin Lowlands, Lake County, Oregon. OSMA Survey Reports 85-5.

Pettigrew, R. M., Paul W. Baxter, and Thomas Connolly. 1985. Archaeological Investigations on the Eastern Shore of Lake Abert Lake County, Oregon, Vol. 1. University of Oregon Anthropological Papers 32.

Pettigrew, R. M. 1980. The Ancient Chewaucanians: More on the Prehistoric Dwellers of Lake Abert, Southwestern Oregon. *Association of Oregon Archaeologists Occasional Papers* 1:49–67.

———. 1979. Archaeological exploration of the proposed improvement of the Pikes Ranch—Valley Falls Section, South Unit, of the Lakeview-Burns Highway, Lake County, Oregon. Oregon State Museum of Anthropology, Univ. of Oregon, Eugene, Oregon. 20 pp.

Pettigrew, R. M., and D. L. Cole. 1977. Archaeological survey of the proposed improvement of the Peterson Ranch-Lake County Line Section of the Lakeview-Burns Highway, Harney County, Oregon Museum of Natural History Report 77-4, Eugene, Oregon. 12 pp.

Phillips, K. N., and A. S. Van Denburgh. 1971. Hydrology and geochemistry of Abert, Summer, and Goose Lakes and other closed-basin lakes in south-central Oregon. Geological Survey Professional Paper 502-B. U.S. Government Printing Office, Washington, D.C. 88 pp.

Reheis, M. 2015. Owens (Dry) Lake, California: A Human-Induced Dust Problem U. S. Geological Survey Article and *Journal of Geophysical Research*. In press.

Reid, J. S., R. G. Flocchini, T. A. Cahill, R. S. Ruth, and R. S. Salagado. 1994. Local meteorological transport and source aerosol characteristic of late autumn Owens Lake (dry) dust storms. *Environ.* 28:1699–1709.

Rubega, M. A., and J. A Robinson. 1996. Water salinization and shorebirds emerging issues. *International Water Studies* 9:45–54.

Schmidt-Nielsen, K. 1960. The Salt Secreting Gland of Marine Birds. *Circulation* 21:955–967.

Stern, M. A., K. A. Kristensen, and J. F. Morawski. 1990. Investigations of snowy plovers at Abert Lake, Lake Co., Oregon. Final Report for Oregon Dept. Fish and Wild. 14 pp.

Stoechkenius, W. 1976. The purple membrane of salt-loving bacteria. *Scientific American* 234:38.

Sun, D. Y., J. Z. Guo, H. A. Hartman, H. Uno, and L. E. Hokin. 1992. Differential Expression of the Alpha and Beta Mesenger RNAs of Na,K-ATPase in Developing Brine Shrimp as measured by In Situ Hybridization. *J. Histochem and Cytochem.* 40:555–56.

———. 1991. Na, K-ATPase expression in the Developing Brine Shrimp Artemia. Immunochemical Localization of the alpha- Beta-subunits. *J. Histochem and Cytochem.* 39:1455–1460.

Tinniswood, W. R. 2007. Adfluvial Life History of Redband Trout in the Chewaucan and Goose Lake Basins. *Redband Trout: Resilience and Challenge in a Changing Landscape.* Oregon Chapter, American Fisheries Society, pp. 99–112.

VanDenburgh, A. S. 1975. Solute balance of Abert and Summer Lakes, south-central Oregon. Geological Survey Professional Paper 502-C. U.S. Government Printing Office, Washington, D.C. 25 pp.

Warner, A. T. Mac Rae, and J. S. Bagshaw. 1987. *Cell and Molecular Biology of Artemia Development.* New York: Plenum Press.

Newspaper, Magazine, and Internet Articles

Davis, Rob. 2015. "Saltwater Lake is Disappearing and Scientists Don't Know Why." Portland, Oregon: *The Oregonian.*

Montana, Cale. 2000. "Klamath Tribe near remedies over disturbed of ancestral remains." Published in INDIAN COUNTRY TODAY MEDIA NETWORK.Com. 5 October 2000, pp. 1–4.

Polansek, T. 2015. Glowing in the dark, GMO chickens shed light on bird flu fight. *Reuters,* http://reut.rs/1EMZTPI

Schmidling, Helen. "Prehistoric Life at Lake Abert." *The Nugget*, August, 2015.

www.ingramcontent.com/pod-product-compliance
Lightning Source LLC
Chambersburg PA
CBHW041058180526
45172CB00001B/16